Solitaire

The Dodo of Rodrigues Island

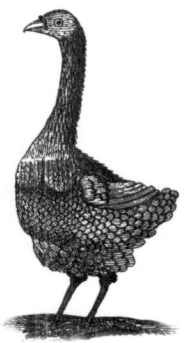

Solitaire
The Dodo of Rodrigues Island

Alan Grihault

Editor: Elizabeth Weaver
Design and layout: Jungle Blue, Mauritius
Printing and binding: Précigraph, Mauritius

Solitaire - the Dodo of Rodrigues Island
ISBN 978-99903-36-30-6

National Library (Republic of Mauritius) Cataloguing-in-Publication Data

Grihault, Alan
Solitaire : the Dodo of Rodrigues Island / Alan Grihault. – [S.l. : s.n.], 2007.
132 p. : ill. ; 24 cm.
Includes bibliographical references and index.
ISBN 978-99903-36-30-6
1. Extinct birds – Mauritius.
2. Solitaire (Bird)
I. Title.
598.65   dc22   07-0009

Cover design: Jean-François Sookahet

To my late mother, Winifred (Nurse Grihault);
my dear wife, Sara;
my children, Nicky, Justin, Steve and Suzy;
and to my good friend, Ralfe Whistler.

Acknowledgements

Photography - Bob Latimer (Latiscope, Mauritius)
As well as being an excellent photographer and a good friend, Bob has given his support to all my various projects.

Editor - Elizabeth Weaver
Elizabeth is an editor in a million and I thank my lucky stars I found her.

Book design - Sandeep Baguant and Jean-François Sookahet of Jungle Blue, Mauritius
Just open the book and you can see what they have done.

Artists
The quality of the book owes a great deal to the following artists who have allowed me to use their work: Vaco Baissac, George André Camille, Fabrice Desvaux de Marigny, Julian Pender Hume, Eric Kwet-On, Bob Latimer, Kathleen Latimer, Leslie Nimmo, Jean-François Sookahet.

Institutions
Many thanks for help rendered by the staff of the following institutions: Alexandra House School (Mauritius), Ambergris NZ Ltd (New Zealand), Artisanat La Colombe (Rodrigues), Blue Penny Museum (Mauritius), CareCo (Rodrigues), Carnegie Library (Mauritius), Casela Nature and Leisure Park (Mauritius), Coin François Leguat Hotel (Rodrigues), Cotton Bay Hotel (Rodrigues), Dodo Research Programme (Leiden, Holland), First Fleet Reproductions Ltd (Mauritius), Forestry Department (Rodrigues), Grande Montange and Anse Quitor nature reserves (Rodrigues), Grande Montagne Visitor and Information Centre (Rodrigues), Hastings Museum and Art Gallery (England), Imprimerie & Papeterie Commerciale (Mauritius), Mauritian Wildlife Foundation (Mauritius and Rodrigues), Mauritius Museums Council (Mauritius), National Heritage Trust Fund (Mauritius), National Library (Mauritius), Philatelic Museum (Mauritius), La Vanille Réserve des Mascareignes (Mauritius), York Museums Trust (England).

Individual acknowledgements
The following kind people have helped me in one way or another; some have helped a lot, some have helped a little, but I would like to thank them all equally: Aurèle Anquetil André, Nelly Ardill, Frans Beuse, Anthony Cheke, Dr Nik Cole, Michel Coquet, Paul Draper, Andy Frost, Errol Fuller, Owen Griffiths, Sara Grihault, Philippe la Hausse de Lalouvière, Julian Pender Hume, Jane and Pierre Lagesse, Lorraine Lagesse, Marko Laine, Bob Latimer, Greg Middleton, Françoise Monpays, North-Coombes family, Jolyon Parish, Nurveen Ratty, Ian Ridpath, Kenneth Rijsdijk, Guy Rouillard, Fred and Debbie Stone, Felice Stoppa, Vikash Tatayah, Rameshwar Tupsy, Arturo Valledor de Lozoya, Gavin Weaver, Leslie Wilcox.

Mauritian business support
Mauritius Tourism Promotion Authority (MTPA)
François Leguat Giant Tortoise and Cave Reserve
AVIS Rent a Car (Rodrigues)
DHL (Mauritius)
Mauritius Glass Gallery (Hands of Fame)
E. C. Oxenham & Co. Ltd.

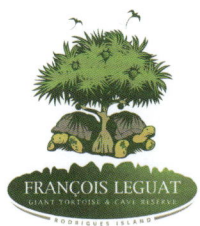

Contents

...the sad fate of the Solitaire is

particularly regrettable, for though it had not the

striking ugliness of the Dodo, which differs from all

other large birds in being developed in the direction of bulk

rather than length in all parts of its body, the Solitaire was

evidently a much more interesting creature.

(Kinn, 1904)

Prologue

The Solitaire was a magnificent bird – not so much for the colour of its plumage as for its size and stately demeanour.

Everybody has heard of the Dodo, but very few know anything about the Solitaire, despite the fact that the two birds were exactly the same species 26 million years ago. Both the Mauritian Dodo[1] and the Rodrigues Solitaire were perfect examples of how birds can adapt to an island lifestyle, easily rising to the top of the pecking order and reigning supreme in a habitat free of predators. They were both the avian equivalent of the pig or deer, eating everything that lay on the ground before them.

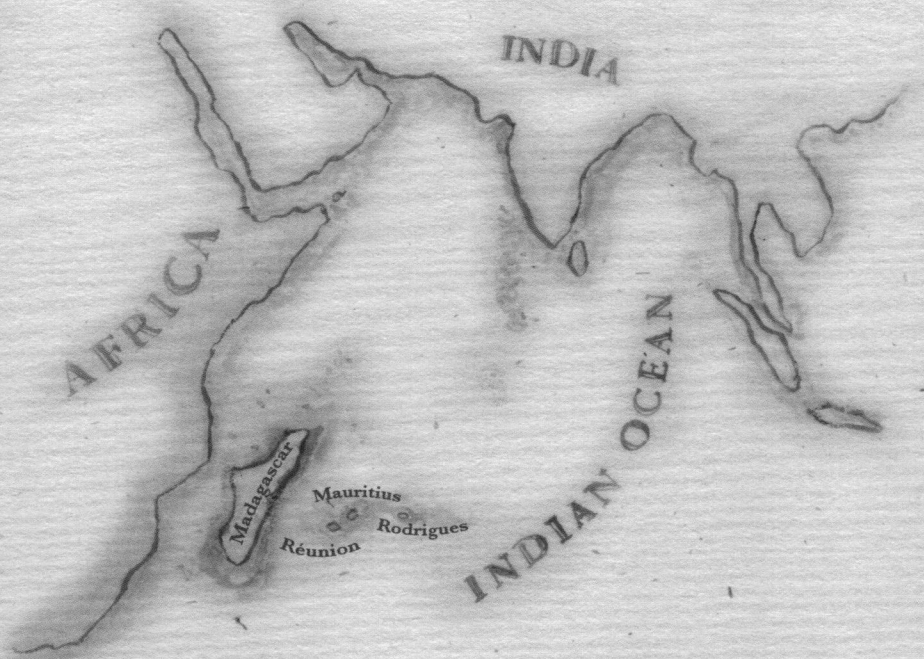

Mauritius, Réunion and Rodrigues (the Mascarene Islands), on the western edge of the Indian Ocean, remained uninhabited by man until the 17th century, presumably because they were not on a major sea route. The Solitaire lived on the tiny island of Rodrigues, which lies 574 km to the east of Mauritius.

Left: The Solitaire. Painting by Julian Hume.

---

[1] All scientific names of plants and animals have been moved to *Appendix A* for more fluent reading.

The first clue that there was a large bird living on Rodrigues came from Sir Thomas Herbert (1638) who mentioned in his *Travels into Divers Parts of Asia and Africa* that the Dodo is *'generated here, and here only* [meaning Mauritius], *and in Dygarroys* [Rodrigues]'.

This Dodo-like, large bird, which we now call the Solitaire, was first described by François Leguat de la Fougère (1708) who, together with his companions, lived on the island from 1691 to 1693. He wrote[2] fondly of the birds and commented that they walked about *'with such pride and good grace that one cannot help but admire and like them, to the extent that quite often their good appearance has saved their lives.'*

They were next described by a naval officer, Julien Tafforet, when he was marooned on Rodrigues for nine months between 1725 and 1726. He recalled that they could be seen *'strutting proudly about, either alone or in pairs, they preen their plumage or fur with their beak and keep themselves very clean.'* Tafforet has often been credited with writing the 'anonymous' document known as the *Relation de l'isle Rodrigue*, and this has been assumed in this book.

From these reports there is no doubt that the Solitaire was an elegant and striking bird, and perhaps more impressive than the Dodo of Mauritius.

After its extinction, the Solitaire failed to command the same interest as the Dodo for various reasons. To start with, the documents by Tafforet were wrongly filed and were lost until a Mauritian magistrate, F. J. J. Rouillard, accidentally found them in the Paris Ministère de la Marine in 1874. This meant that for a long time the description of the Solitaire came from just one eyewitness.

The promotion of the Solitaire was not helped by the fact that many scientists cast doubt on the credibility of Leguat, and even those who accepted his accounts thought that it was the same bird as the Dodo of Mauritius. J. V. Thompson (1829), a surgeon and zoologist who read Leguat's book, notes that *'we have in this last relation of Leguat, who resided amongst them for a considerable period, a detailed, although rude, description, and a natural history of the Dodo, probably the only one that was ever penned under such favourable circumstances.'*

The editor of Thompson's work then adds that *'it is not likely that the three islands of the Mauritius group possessed each a distinct type of so singular and unique a bird.'*

This failure to recognise the Dodo and the Solitaire as two separate species was completed when Georges Cuvier (1830), after receiving some Solitaire bones from Rodrigues, noted that they were found

---

[2] Many of the original documents quoted here were originally written in French. For the purposes of this book English translations have been used.

'*under a bed of lava, and in Mauritius*'. This mistake was unforgivable, as no Dodo bones had been found in Mauritius at that time.

One of the aims of this book is to inform readers about the natural history of the Solitaire, which found a small niche for itself on Rodrigues for hundreds of thousands of years until it finally became extinct. It is hoped that an awareness of this bird, the island where it lived, and its demise will promote an understanding of the importance of conserving our environment and its dependent biodiversity. This in turn may encourage individuals, schools and local communities to become involved in conservation projects, to help rescue Rodrigues from being one of the most environmentally degraded tropical islands in the world and turn it into a conservation success.

A view of Rodrigues.

1 The island of Rodrigues

Rodrigues can be quite rightly thought of as just a hill in the sea, but, in spite of its small size and low-altitude mountains, it is an island of twisting and tortuous roads that go up and down steep mountainsides.

Rodrigues is an isolated island lying east of Mauritius and measuring a mere 18 km long and 8 km wide. It can boast only one main ridge, which is about 11 km long, rising to a peak at Mont Limon, 398 m above sea level. The emergence of this volcanic ridge has not yet been precisely dated, but estimates of its age vary from 1.5 to 4 million years, and it has even been suggested that it may be older than Mauritius, which emerged from the sea nearly 8 million years ago. The ridge is cut into deep ravines opening into valleys, which run down to the sea, leaving little room for any flat land. It is the only Mascarene island with extensive limestone deposits, and is surrounded by a large fringing reef enclosing eighteen small islets. When it was first discovered in the late 1400s or early 1500s, it was entirely covered with evergreen, palm-rich forests.

Above: L'Hermitage Island from Rodrigues with the reef in the distance.
Left: Pandanus growing on the side of a rocky valley. Photograph by Bob Latimer.

Geologically, the island is very similar to Mauritius, of basaltic origin; but a large portion of the south-western part is composed of very ancient up heaved coral, abounding in fissures and caverns, large and small. (Caldwell, 1875)

Arabian explorers and traders knew of the three Mascarene islands of Rodrigues, Mauritius and Réunion. This knowledge was passed on to the earliest Portuguese maps (Cantino, 1502), which were developed after the Portuguese explorer, Bartholomew Diaz, rounded the Cape of Good Hope in 1488. The islands were named *Dina Moraze*, now Rodrigues; *Dina a Robi*, now Mauritius (from the Arabic *Diva harab*, meaning desert or square island); and *Dina Margabin*, now Réunion. In later maps, Rodrigues is called *Diogo Roiz* (*Roiz* being the diminutive of Rodriguez), which leads to the assumption that it was named after Diogo Rodriguez, the pilot of the *Albuquerque* (part of a fleet of ships making its way back across the Indian Ocean in 1528, under the command of Pero Mascarenhas). See *Appendix B* for names given to Rodrigues throughout history.

The Portuguese showed no real interest in the uninhabited islands of the Indian Ocean, as their main interest was in converting natives to Christianity, and there was also the chance of making much profit by attacking the Arabs and taking over their trade. They soon built up trade on their way from the Cape to India, keeping fairly near to the coast, and missing out the islands, unless forced to visit them when they were blown off course by storms. There is some uncertainty about Portuguese events at this time, due to the fact that there was an earthquake in Lisbon in 1707, causing fires that ultimately destroyed most of the data stored in the archives.

Stamp showing Cantino's 1502 map.

The Dutch recorded that they landed in Rodrigues in September 1601, under the command of Admiral Wolfert Harmensen. The flagship for this expedition was the *Gelderland*, which was one of the original ships to call on Mauritius with Vice-Admiral Warwyck in 1598. They reported that it was difficult to find an anchorage and there was very little fresh water, but they did find plenty of food and water after passing through a narrow gap in the reef on the north side of the island (later to become Port Mathurin, the capital). After this short stay they decided to make sail for Mauritius.

Nothing else was reported about Rodrigues until the Dutch admiral Pieter Wilhem Verhoeven passed by 'San Roderigo' in 1611, on his way back, via Mauritius, to the Netherlands, but no landing was mentioned.

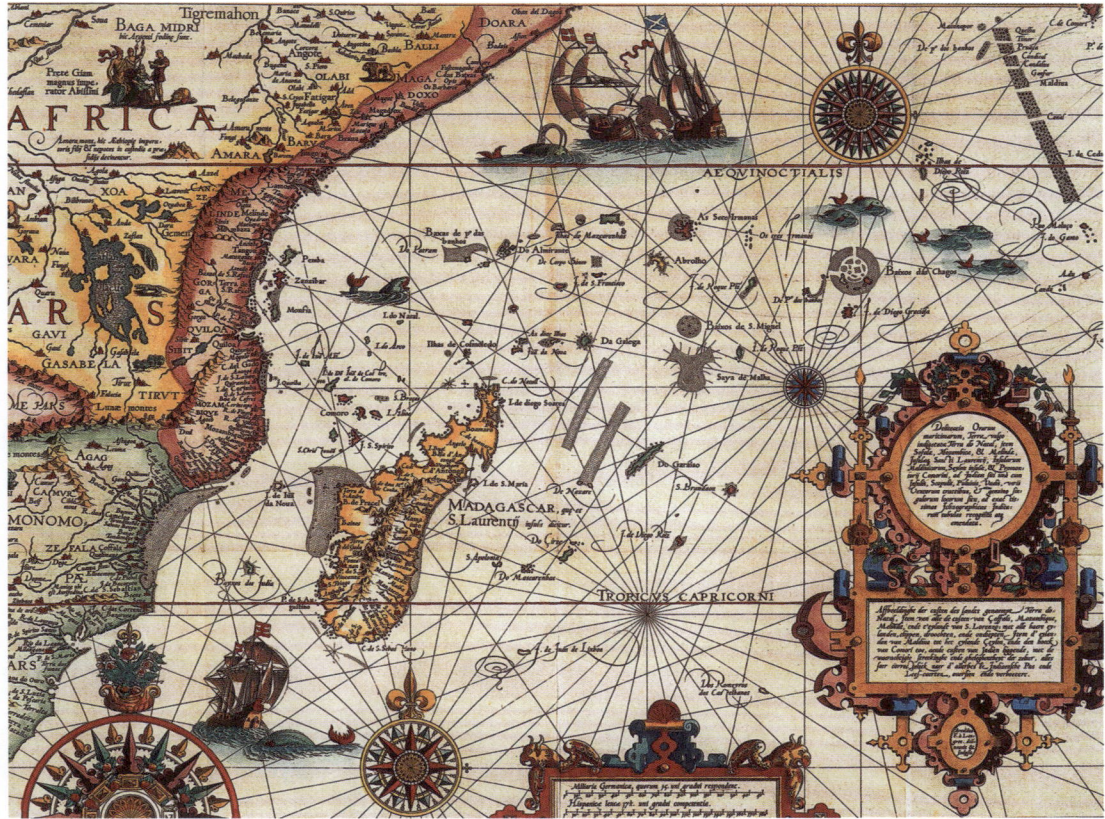

Map of the Indian Ocean by van Linschoten and van Langren (1597), courtesy of IPC Ltd and Blue Penny Museum.

Then, in 1639, Adriaan van der Stel was instructed to stop in at Rodrigues, on his way to becoming the commander in Mauritius, in order to check an earlier report that ebony was easier to obtain there. Later, he was asked to take possession of the islands close to Mauritius and to leave some mark to show that this had taken place, but bad weather stopped him from doing so.

The first temporary settlement took place in 1644 when the Dutch ship *Bereckout*, on its way from Batavia in Java (now Jakarta) to Mauritius, anchored off Rodrigues. A party of about twelve men set off in their long boat to land on the island, but soon afterwards the weather deteriorated. Having waited for two days offshore, the *Bereckout* set off for Mauritius without the landing party. After three months in Rodrigues, the abandoned men decided to sail for Mauritius in their long boat. One of the men, Gerrit Andriesz (the ship's helmsman), gave a verbal report that Rodrigues contained no ebony or anything else of value to the Dutch East India Company (Heeringa, 1895), so the company lost all interest in the island.

Much later, in 1691, a party of French Protestant refugees arrived on the island. At this time many Huguenots were escaping from the persecution by Louis XIV in Catholic France, and this group had been attracted by the promise that they would help to form a republic of Protestants on a distant island they would call 'Eden'. In the end only eight colonists, under the leadership of François Leguat, arrived to take possession of Rodrigues in the *Hirondelle*. They had hoped to settle in Réunion but, as the French already occupied this island, they were forced to sail for the alternative, smaller island.

change of plan, since François Leguat left us a legacy of what the island of Rodrigues was like in the late 1600s, as well as accurate descriptions of the flora and fauna of the island.

Leguat and his companions spent two years on the island before they built a raft and sailed to Mauritius, where they were accused of being spies and trying to steal precious ambergris. Finally, after spending a miserable three years on a barren, rocky island off Mauritius, Leguat found his way to England through Batavia. He then published his journal in 1708, which in turn alerted the Admiralty that it would be worth their while to survey the island.

François Leguat de la Fougère (1638-1735)
François was the son of Pierre Leguat, born in the small east-central French province of Bresse, four or five years before the deaths of Cardinal Richelieu and King Louis XIII, in 1642 and 1643 respectively. He was a farmer and a Huguenot (Protestant). He was driven into religious exile during the reign of Louis XIV, who made it a crime to stay in France if you were a Protestant. He reached the Netherlands in 1689, but soon after left on a project that was to settle on what they thought would be the empty island of Réunion. Instead, they went to Rodrigues because Réunion was already inhabited by the French.

After two years on Rodrigues they sailed to Mauritius, where from 1693 to 1696 they were imprisoned by the Dutch commander Roelof Deodati. They were then sent to Batavia where they were still confined, but they finally arrived on Dutch soil in 1698. Soon after, Leguat made his way to England where Protestants were welcomed under the rule of William of Orange. Leguat published his books in

---

Courtesy of www.ambergris.co.nz. Photograph by Frans Beuse.

Ambergris
A waxy substance consisting mainly of cholesterol secreted by the intestinal tract of the sperm whale and often found floating in the sea, used in the manufacture of perfumes. [Old French *ambre gris*, meaning grey amber.]

For the sake of history, and the study of the Solitaire in particular, it was a fortunate

The French were also alerted, but neither country took any serious action until the

French sent thirty-eight colonists (ten soldiers, thirteen farmers and fifteen slaves) from Réunion to the island, on the ship *La Ressource* in September 1725. The captain reached Rodrigues but found it difficult to find the entrance to Port Mathurin. It was decided to lower the boat and the second mate, Julien Tafforet, and four men were instructed to find a passage between the reefs. Meanwhile, a sudden squall caused the ship to lose her anchor. The ship was compelled to turn back to Réunion, and could not return to Rodrigues, as it was nearing the cyclone season. The group had to survive on Rodrigues for nine months until they were rescued in June 1726. During his enforced stay Tafforet surveyed the island and produced his *Relation de l'isle Rodrigue*, in which he described the Solitaire.

One of the main resources of the island was the abundance of tortoises, which in some places stretched as far as the eye could see. This unfortunate animal could be loaded

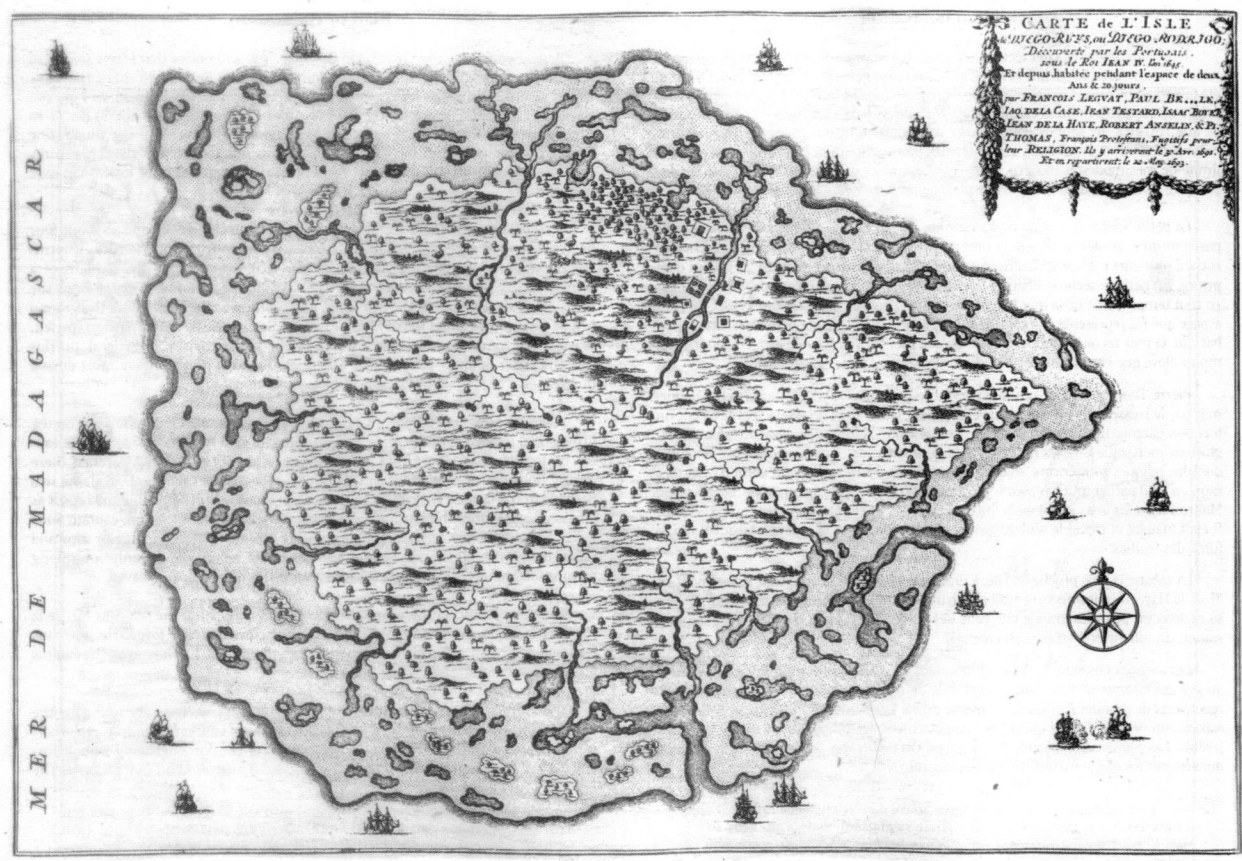

Map of Rodrigues by François Leguat from his *Voyage et Avantures* (1708).

onto ships and kept alive for long periods, supplying fresh meat for the whole of the voyage. The first organised establishment for the gathering and exportation of these animals seems to have been ordered by Mahé de Labourdonnais after he became governor of Mauritius. By 1736, there are records to show that small enclosures were constructed by a few soldiers, lascars and slaves, and the ground was cleared to make a kitchen garden. It has been estimated that at least 10,000 tortoises were exported per year during the time of Labourdonnais, and

his successor records that over 17,000 tortoises were placed on ships in a two-year period from 1759 to 1761.

Little is recorded about the fate of the Solitaire during this mass exportation of tortoises. In 1761, a French expedition went to Rodrigues to make astronomical observations on the passing of the planet Venus across the sun's disc. A member of this scientific team was a Catholic priest who was also a distinguished mathematician and astronomer, Abbé Alexandre Guy

Le Duyfken – an early visitor to Rodrigues. Painting on a wall at the Centre for Human Resources, Malabar (Rodrigues).

Latimer

Pingré. He had read Leguat's journal and used it as a roadmap to survey the island, and actually positioned his observatory on a small hill overlooking Leguat's original settlement. He also found time to look for Solitaires, as he had been told that there were still a few about:

> *By the time of Pingré's visit the bird had virtually disappeared; one or two could perhaps have been seen in the most remote parts of the island, but Pingré did not see any. Soon afterwards, Solitaires had become extinct: only their remains would now excite man's interest and draw attention to their island.*   (North-Coombes, 1971)

Abbé Alexandre Guy Pingré (1711-1796)
Alexandre was born in Paris and was educated in Senlis at the College of the Genovefan Fathers, Regulars of the Order of St Augustine, which he entered at sixteen. In 1735, he was made professor of theology there. Around 1749, he accepted the professorship of astronomy in the newly founded academy at Rouen. Later, he was made librarian of Ste-Geneviève and built an observatory at the abbey. In 1753, he compiled the first nautical almanac. He joined an unsuccessful expedition to Rodrigues to observe the transit of Venus in 1761, when he spent four months on the island. More satisfactory results were obtained from an expedition to the French Cape on Haiti, where the next transit was observed in 1769.

Pingré seems to be the last of the visitors to make any mention of the Solitaire; a bird that perhaps had the ability to make itself scarce at the approach of man, unlike the slow and cumbersome tortoise. Martenne de Puvigné, commandant of the tortoise establishment at Rodrigues (1751–1762), told Pingré that the Solitaire was found only in remote corners of the island, and this recorded statement was the last mention of the bird.

Meanwhile, the tortoises were disappearing fast and the small island of Rodrigues was abandoned, as it was not much use to anyone without them. The population of around 300,000 tortoises had gone forever, with the last reported sighting *circa* 1795.

From as early as 1794, a British garrison, under the command of Lieutenant-Colonel Henry S. Keating, used Rodrigues as a base for the British blockading squadrons off Mauritius and Réunion. Ships and troops were assembled in Rodrigues in order to attack Mauritius and Réunion, and both islands were taken in the year 1810, resulting in some British forces remaining in Rodrigues until April 1812.

By this time the Solitaire had long disappeared, so it is necessary for us to concentrate our attention on the time when it was still around.

Pandanus fruit

From an original photograph by Bob Latimer.

2 The ancient forest

As fine a country as a man can desire to beeholde, although woody. (Mundy, 1628)

The detailed accounts of Leguat from the early 1690s paint an idyllic picture of *'valleys covered with palm-trees, lataniers, ebony's and several other sorts of trees'* and *'rivulets of fresh water, whose springs are never dry'*. Other early travellers to the Mascarene Islands were impressed by the forests they observed, so much so that Captain Castleton described Réunion in 1613 as *'England's Forest'* (Tatton, 1625). Peter Mundy made similar descriptions of Rodrigues, during his visit in 1628.

Even in 1809, when some British forces landed in Rodrigues for their sea offensive against Mauritius, the forests were still ample in dimension. Keating boasted that *'we have timber nearly of every kind within our reach and in a few weeks we shall be able to wood and water all ships touching here with great facility and expedition.'*

Leguat described some of the trees and plants, but admitted that he was more interested in the fauna than the flora of the island. One tree that impressed him was the Indian species of the banyan tree, which offered shade to the settlers while its *'thick branches extend in a circle so that the rays of the sun find it impossible to penetrate, and some are so big that two to three hundred people could find shelter under it.'*

Often called 'the many-footed tree', because of the aerial roots striding from its trunk to form a small forest, Leguat tells us that the French called this a Banian tree because the Indian merchants, or Banias, conducted their business in its shade. In the many rural areas of India, the banyan is still used for council by village elders and for religious worship by priests. Leguat's further description of the tree shows his eye for detail:

> *The Rodrigo Kastas (for I ought to keep the Indian name at least in the Indies) bear leaves as broad as one's hand, pretty thick, and somewhat like that of a lilac or a heart in shape, they are softer than satin to touch. Their flower is white, and smells well: their fruit is red and round, and as big as a black Damask plum. Their skin is hard, and within it is a thin seed, a little like that of a fig. The fruit is not prejudicial to health, but 'tis insipid. The bats commonly feed upon it, and multitudes of them nest in the tufted branches of this tree.*

Leguat also describes the Stinkwood tree, which according to Balfour (1879) is probably the tree referred to by Leguat as the Nasty tree or the Bois Cabris of the Creoles. This small tree is easily recognised by its disagreeable odour, which accounts for its popular name. Some have thought that Leguat referred to the Bois Puant, but the odour of that tree, though exceedingly objectionable, is only apparent when the sun shines upon it, and is not like the persistent odour from the Bois Cabris. The foetid smell comes from pungent oils, which prevent the timber from rotting or being attacked by termites.

A young Yellow Latanier plant.

Another tree which Leguat described in detail was the Pavillon: a Rodriguan species of pandanus or screw pine:

Among the numerous, diverse trees nature planted here, there is an admirable one which is particularly noteworthy for its beauty, height, roundness, and the rare symmetry of its magnificent branches. All the ends of the boughs are greatly tufted with the thick foliage falling all around almost to the ground, so that from whichever side one approaches this handsome tree only a small portion of its trunk may be seen – sometimes none at all. The centre of this tree is as shaded as one can imagine, and the interior branches inside resemble wooden supports a carpenter has set there to hold up the tufts which surround it, thus making the interior a kind of cage or tent. Truly, the outside of this tent is most beautiful and charming, but the shade and freshness of the interior also have their delights.

Unfortunately, the fruit of this marvellous tree is not good to eat. Those of us that have tasted it out of curiosity have found it bitter but we know from experience that it is not dangerous. It smells very much like a ripe quince and looks like a sort of cluster with compact seeds; from a distance it looks like a pineapple although there is a marked difference between these two plants. As for me, I wanted to call it 'Pavillon'. The leaves are green and the leaf stems so short that they seem directly attached to the branches. The largest are four or five inches wide at the top, ending in a point; their length is about fifteen inches. They form large bouquets interspersed here and there with clusters differing in colour, depending on their ripeness. I have often walked around this natural palace, always thrilled equally with its size and unique beauty.

A tree that proved very useful to Leguat and his companions was a species of latanier palm. When they arrived and constructed their huts 'the walls were built of the trunks of the Latanier palm trees, the roofs being covered with the large leaves of the same tree.' Later, when they built their raft to escape from Rodrigues, they 'made diverse kinds of ropes with thread or fibres coming from the stems of leaves of the Latanier trees.' The species found on Rodrigues is called Latanier Jaune (Yellow Latanier) because of the yellow leaves of the young seedlings. Its fruit, which is now used for making rum, must have been enjoyed by the Solitaires and other birds.

Heterophyllous plants

Many plants and trees found on Rodrigues are heterophyllous, meaning that they have more than one type of leaf on the same plant. The young plants have leaves that are a different colour and shape from those that grow later on the adult plant. At one stage in the plant's growth there will be two or more distinct types of leaves on a single individual plant. The reason for this type of development is still unclear, but one theory is that the shape and colour of the young leaves at the base of sapling trees may discourage hungry animals, such as tortoises, from foraging.

Unfortunately, the forests of Rodrigues have suffered badly since man set foot on the island, over 300 years ago. The destructive advent of fires, direct exploitation, clearance for agriculture and the introduction of alien flora and fauna changed Rodrigues rapidly. In 1879, Balfour referred to the island as a *'dry and comparatively barren spot, clothed with a vegetation mainly of social weeds.'* Today, no contiguous areas of native forest remain.

Nevertheless, the importance of its biodiversity is still very great, with 145 native and 49 endemic plants recorded. The flora of Rodrigues also includes six endemic genera. Nine of its endemic species, however, are down to less than ten mature individuals in the wild. In the past, extensive reforestation has been almost exclusively of non-native species, but more endemic trees have been planted in the last few years.

An example of the old policy can be seen by visitors who cannot help but notice that much of the coast is still dominated by the Australasian native *Casuarina* (Filao) trees, which were planted around the coastal regions of Mauritius and Rodrigues in the 1930s. Over the years it was found that these trees do little to bind the sandy soil and in fact leave it acidic and unsuitable for other vegetation to grow, as well as dropping sharp cones which litter the beaches and make it uncomfortable for people to walk or sit. Since 2001, the *Casuarina* trees have gradually been replaced by endemic coastal trees, which have quick-spreading roots. These trees are more resilient to drought and the roots help contain soil sediment during periods of heavy rain.

It will take time, but Rodrigues and its people are slowly getting parts of the island to look something like the 'woody' habitat of the elegant Solitaire.

Present-day landscape.

Latimer

Café Marron in Grande Montagne Nature Reserve. Photograph by Elizabeth Weaver.

The Rodrigues Coffee tree that refused to die Café Marron (*Ramosmania rodriguesii*)

In 1877, a European visitor to Rodrigues made a drawing of a Café Marron plant, which by the middle of the 1900s was thought to be extinct. Then, in 1979, a biology teacher on Rodrigues handed out copies of the 1877 drawing to a group of 12-year-olds, and one of the boys informed him that he had such a plant near his house. The teacher took a sample branch and sent it to the Royal Botanic Gardens, Kew (London), where it was identified as wild coffee. In fact, it was the last wild Rodriguan coffee plant in existence.

The plant was growing at the side of a busy road and a fence was put around it, but this only served to attract attention and local people started taking bits off the tree. More cuttings were sent to Kew Gardens and in 1986 successful root cuttings were made. By 2003, the first fruit was produced on plants with viable seeds. Now a few plants are being reintroduced into Rodrigues, and visitors can see them at Grande Montagne Nature Reserve.

The Pig and the Magic Stone, a digital painting of rural Rodrigues, by Leslie W. Nimmo.

3 The first eyewitnesses

What was Leguat's contribution to natural history? His description of the Solitaire is, of course, unique, and for this alone he should be remembered, as indeed he is.   (North-Coombes, 1979)

The ship *Hirondelle* left Amsterdam on 10 July 1690, under Captain Antoine Valleau, with a group of Huguenot refugees sailing to Réunion to make a new life on the island they called Eden. The group was disappointed not to have landed on Martin Vaz Island (Tristan da Cunha) on their way, but the ship eventually reached the Cape on 26 January 1691. During a three-week stay, they tried to find out whether Réunion was still uninhabited, but soon discovered that the French occupied the island. They set sail for Mauritius in order to get more information, but having lost their direction in a cyclone they passed Mauritius and made their way to Rodrigues.

At last, on a beautiful Saturday morning on the 25th April 1691, we sighted new land. It was the little island of Diogo Roiz where our captain had decided to leave us. We did not at first perceive either port or bay, or any place that we might land. In the evening we sounded and met with the bottom three leagues from the shore. We cast anchor, it was calm, and we stayed there until Monday the 27th, for what reason I know not.   (Leguat, 1708)

Eventually, François Leguat landed on Rodrigues on 1 May 1691. He had come a long way from his birthplace in the French province of Bresse. He had been a farmer and a Protestant who was driven into religious exile by Louis XIV in 1685. By all accounts he was a rather bigoted and narrow-minded fifty-three-year old, but he was a good observer of nature and was able to record his observations with skill, elegance and accuracy.

Model ship similar to the *Hirondelle*, courtesy of First Fleet Reproductions Ltd.

Left: Rodrigues in 1691, from Leguat's *Voyage et Avantures* (1708).

Leguat and his group established themselves on the north part of the island at the mouth of the Grande Rivière, on a site now known as Port Mathurin. Besides Leguat, the group comprised the following men:

Robert Anselin	the son of a miller from Picardy
Isaac Boyer	a merchant; the son of an apothecary from near Nérac
Paul Bénelle	an educated man; the son of a merchant from Metz
Jacques de la Case	an ex-army officer; the son of a merchant from Nérac
Jean de la Haye	a silversmith from Rouen
Jean Testard	a druggist; son of a merchant from St Quentin
Pierre Thomas	a ship's pilot who quarrelled with his captain and chose to remain in Rodrigues.

1 *Pierre Thomas*
2 *Jean de la Haye*
3 *Robert Anselin & the kitchen*
4 *The garden*
5 *François Leguat*
6 *Jacques de la Case*
7 *Jean Testard*
8 *Paul Bénelle and Isaac Boyer*
9 *The big tree where we had food*

Leguat's village. Painting by J.-F. Sookahet, inspired by Leguat's original map.

Seven huts were erected which were constructed from the trunks and leaves of latanier palms, and each was surrounded by a small garden protected by the trunks of the palms. Each colonist had a hut to himself, except for Bénelle and Boyer, who shared. The huts were about 4 m² in size.

The *Hirondelle* remained at anchor for fifteen days while the new settlers unloaded their stores and finished building their small dwellings. The captain and settlers had little opportunity to explore the island, but they observed a great number of tall birds similar to birds found on the island of Réunion. The latter bird was known as the Réunion

Solitaire, and the name Solitaire was also given to the newly discovered birds on Rodrigues. It seems that all the descriptions of the so-called Réunion Solitaire, and perhaps the Réunion White Dodo, actually referred to the extinct Réunion Flightless Ibis (see *chapter 10*).

Whatever the case, when Captain Valleau returned from Rodrigues to the Netherlands in the following year, he recorded seeing Solitaires on Rodrigues (Valleau, 1692).

After settling down, Leguat soon started recording everything around him. Milne-Edwards (1896) noted that Leguat drew a

Present-day view of the river mouth where Leguat and his companions made their settlement.

true picture of the island as it was at the end of the 17th century by examining the plants and animals, and giving very full and interesting details of their characteristics and their uses. Rodrigues was covered in forest, and the surrounding sea provided a rich harvest, so the companions had no problems in finding sufficient food:

> We had an abundance of meat and fish of our choice – roasts, porridge, soups, stews, herbs, roots, excellent melons as well as other fruit, good palm wine, and clean and pure water … we were of good cheer, without distaste, indigestion, or any kind of sickness, thanks to the Lord – and without bread. The captain had left us two large barrels of biscuits, but we used them rarely to make soups, and often did not think of them.

Leguat mentions seeing two to three thousand tortoises all in one group, as well as Dugongs, which appeared in numerous troops in the sea around the island. He also mentions the wonderful shade provided by the banyan trees. It is Leguat's observations of the fauna and flora on Rodrigues that have become so valuable to the world of natural history.

Obviously Leguat was unaware of the recognition he would get in later years, as he minimises his efforts when writing about the occupations of the settlers:

> Our activities during our stay on this island were not very important, as one can imagine, but we had to occupy ourselves. The upkeep of our huts and cultivating our gardens took a good bit of our time. We also walked around the island. We often went to the southern part as we took our walks, and there is no place we did not explore thoroughly. There are neither high mountains nor hills without greenery, although there were a great number of rocks. The subsoil, which is rock, is covered with two, three or even four feet of soil, and even between the stones, where there was no soil, extremely large, tall and straight trees would grow. From afar, this gives an idea of the island that is more lush than it deserves because one would think its soil is excellent everywhere.

After two years on the island, Leguat and his companions decided they had had enough of being isolated. They built a boat and sailed to Mauritius, where they were imprisoned by Commander Roelof Deodati. He accused the group of spying and stealing ambergris, which they had actually brought with them from Rodrigues.

Deodati banished the unfortunate Huguenot refugees to an island off Grand Port on the south-east coast of Mauritius, called Île Vacoas. This small, rocky island is sandwiched between Île de la Passe and Île aux Fouquets, where a lighthouse now stands. These three islands were named the Three Brothers by the Dutch, and are about 5 km offshore. It was tragic that these survivors from the beautiful island of Rodrigues had to endure captivity on what was essentially no more than a barren rock in the middle of the sea. After this ordeal they were sent to Batavia, where they spent more time in prison until the Dutch Council pronounced their innocence. Leguat and two other survivors of the original Rodrigues Huguenots were eventually returned to the Netherlands in 1698. Finally, Leguat sailed to England where he joined many other French Protestant refugees, and published a book of his *Voyage et Avantures* in 1708.

Hermit (2002) gives an interesting fictional glimpse of a vessel that may have been the one constructed by Leguat and his companions to sail from Rodrigues to Mauritius:

A great Canoe had been fastened with timber to a smaller float. Between these hulls a mast had been erected, upon which a lateen like sail was positioned. The vessel had no rudder, but was steered with an oar. It carried no ballast, and had no keel, relying only on the resistance created by the hulls for stability. [They] took the vessel into the bay, and discovered that it could sail faster than any other boat ... Unfortunately it was too small to be useful for any commercial purpose, and its rapid up and down bobbing made it anything but comfortable.

Hermit (2002) also fictionalises a visit to Rodrigues some years after Leguat and his companions had left the island, and his description gives us an insight into how the island might have looked:

The next day we spent on Rodrigues, which has no natural harbour, although a most beautiful coral reef surrounds the island. Having passed the reef, we found that while the island has a black volcanic cliff all the way around it, this cliff is pierced with many delightful bays. Each one of these is a gem of nature, with clear sea water, having scarcely a ripple running up to a small beach invariably headed by a cluster of palms standing sentry duty. On the north-north-west coast, we finally found Leguat's abandoned settlement, where a stream ran down a small, well-wooded valley, ending in a marshy area; but the lack of a harbour explained why there had been so little interest in the colonists. These settlers had built crude houses of latanier trunks, but they were already collapsing back into their natural state, and their garden had been devastated. In the remnants we found a giant among tortoises, fully 6 feet across. We slaughtered it, and sent it back to the ship with a party of sailors.

From this tortoise we obtained over three hundred and fifty pounds of meat. The settlers had left a record of their occupation inside a bottle, placed in a niche they had carved in a tree. We made a fair copy and witnessed it, adding as a codicil that of their subsequent history with which we were familiar. We removed the original to take as evidence to Batavia, but placed the copy back into the receptacle as a memorial to these brave (or stupid) souls. While exploring the area, we found a gigantic cavern, quite filled with stalactites and stalagmites, to say nothing of delicate waterfalls of crystal, with a stream running through the centre. A little further up the cave I found a dry area, accessible only via a natural construction that looked like a crystal pulpit, in which skeletons of strange animals and birds were scattered. We looked for signs of a predator, but finding nothing, assumed that this area must have fulfilled a similar function for these creatures as the legendary elephant graveyards in Africa.

4 Who were the Solitaire's parents?

When the Rodrigues Solitaire was recognised as belonging to the pigeon family, scientists believed that, like the Dodo, it probably evolved from a fruit pigeon of the Treronidae family, which reached the Mascarenes from Madagascar or Africa.

Recent DNA evidence (Shapiro *et al.*, 2002) has shown that the Dodo and the Solitaire indeed had a common ancestry with the Treronidae family, but actually descended from the Nicobar Pigeon. The two birds ended up on two different islands, with the Mauritian Dodo becoming famous as a symbol for extinction, while its cousin, the Rodrigues Solitaire, remained almost unknown.

Above: Tooth-billed Pigeon from Owen's *Memoir on the Dodo* (1866). Courtesy of IPC Ltd.
Left: Wood-cut print of a Tooth-billed Pigeon, from Wood's *Illustrated Natural History* (1897). Courtesy of IPC Ltd.

In 2002, scientists from Oxford University's Department of Zoology and the Natural History Museum, London, examined tissue found attached to a foot belonging to a Mauritian Dodo, which has been exhibited in various museums in Oxford since 1683. The team managed to extract sufficient soft tissue from inside a claw of this foot to test and compare it with the DNA of other birds, including the Solitaire. This enabled them to build a family tree for the Solitaire.

The DNA survives. It's very damaged and broken down into tiny pieces, but little fragments remain in the cells preserved in the old specimens.

(Cooper in BBC News, 2002)

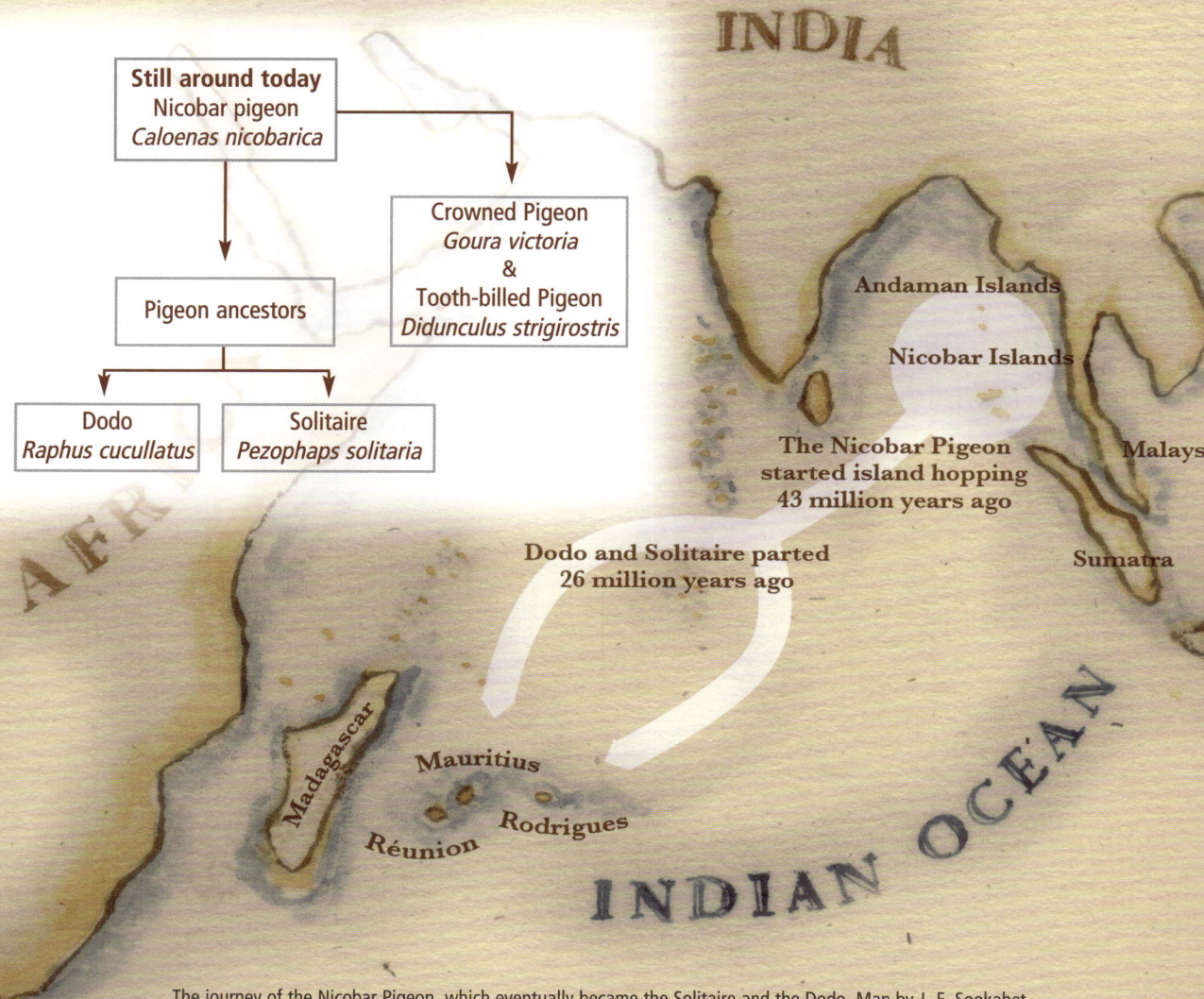

The journey of the Nicobar Pigeon, which eventually became the Solitaire and the Dodo. Map by J.-F. Sookahet.

After analysing the DNA material, the scientists agreed with Reinhardt (1842) and Strickland and Melville (1848) that the Dodo and the Solitaire were both pigeons:

> *Despite the substantial morphological differences between the Dodo, the Solitaire, and other pigeons, the ML analysis shows them to be nested with the family Columbidae.*   (Shapiro *et al.*, 2002)

The earlier suggestions that the Solitaire and Dodo were related to the Crowned Pigeon and Tooth-billed Pigeon were fairly accurate, but the DNA tests have revealed that the closest living common relative is the Nicobar Pigeon, from the Nicobar and Andaman islands (situated 1300 km east of Sri Lanka, and 5000 km from Mauritius). Nicobar Pigeons are nomadic and commute from island to island in flocks of up to eighty-five birds. They can be seen feeding off the ground, tossing leaves aside and digging with their bills. These

birds are usually silent, but they occasionally vocalise during the breeding season when the male bows to the female and coos loudly.

The DNA results point to the fact that the Solitaire and Dodo diverged from their ancestral pigeon stock about 43 million years ago, which means that they existed long before the Mascarene Islands erupted out of the sea. It has been suggested that these birds used mountain ridges, which were above sea level at the time, as stepping-stones, and finally reached the Mascarene Islands. Evidence shows (Shapiro *et al.*, 2002) that the Solitaire and the Dodo were closely related, and parted company about 26 million years ago, ending up on their own islands to become distinct species.

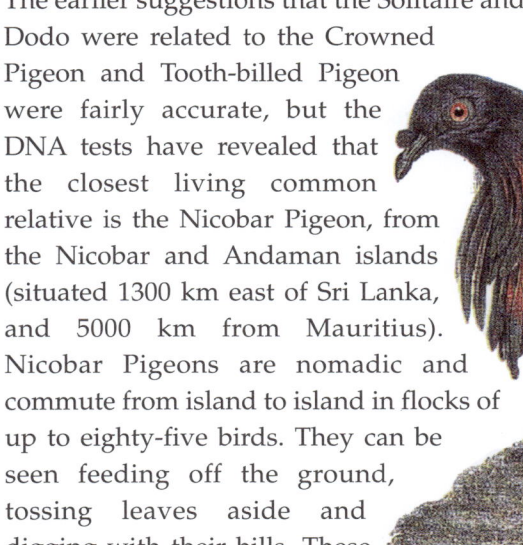

Nicobar Pigeon painted by Pauline Knip in *Les Pigeons* (1811).

The Columbiformes

The Solitaire has been classified into a large group of birds called the Columbiformes, under which pigeons and doves[3] are listed:

The order of Columbiformes				
Suborder	Family	Subfamily	Genus and species	Common names
Pterocletes	Pteroclidæ			Sandgrouse
Columbæ	Raphidæ	Raphinæ	*Raphus cucullatus*	Mauritian Dodo
	Pezophapidæ	**Pezophapinæ**	***Pezophaps solitaria***	**Rodrigues Solitaire**
	Columbidæ	Gourinæ		Crowned Pigeon
		Didunculinæ		Tooth-billed Pigeon
		Columbinæ		Pigeon, Dove, Ground Pigeon
		Treroninæ		Fruit Pigeon

Giving a name

Names often describe the characteristics of living things: how they look, how they move, how they behave, and how they remind us of other similar living things. A number of names for the Solitaire have been reported throughout history, and most of them are shown here (Hachisuka, 1953):

Present name:

> **Rodrigues Solitaire** *Pezophaps solitaria* (Gmelin, 1789).

Previous names:

> *Didus ineptus* – Newton A. in *Ibis*, 1865

> *Didus nazarenus* – Bartlett in *Proc. Zool. Soc.*, 1851

> *Didus solitarius* – Gmelin, 1785

> **Dodo** – Cuvier in *Edinburgh Journ. Nat. and Geogr. Soc.*, 1831

> **Dronte** – Cuvier in *Ann. Soc. Nat.*, 1830

> *Pezophaps minor* – Strickland in *Contr. to Ornith.*, 1852

> *Pezophaps solitaria* – Strickland, 1848

> *Pezophaps solitarius* – Newton A. in *Encyclopaedia Britannica*, 1875

> **Solitaire** – Leguat, 1708

> **Solitary Dodo** – Latham in *Gen. Syn.*, 1785.

---

[3] *Columba* is Latin for dove or pigeon. There is no difference between the two except that the smaller varieties are usually called doves.

Strickland proposed *Pezophaps* (from the Greek for pedestrian and pigeon), hoping that future discoveries of bones would prove him right. He maintained that the Solitaire was a pigeon because of the following characteristics observed by Leguat:

- Fed on dates and plantains
- Had monogamous habits (having only one mate)
- Laid only one egg
- Built a nest
- Inability of nestling to fend for itself.

So, finally the Solitaire was given a scientific classification:

Phylum	Chordata
Class	Aves
Order	Columbiformes
Family	Raphidae
Genus	*Pezophaps* (Strickland, 1848)
Species	*solitaria* (Gmelin, 1789).

Several other birds have been named Solitaire, and a list of these is given in *Appendix C*, along with the name of the Solitaire in different languages.

Nicobar Pigeon at Casela Nature and Leisure Park, Mauritius.

5 The elegant bird

The Solitary Bird mentioned by Leguat and Carré, and the Bird of Nazareth by Father Cauche, seem to bear a great resemblance to the Dodo, though they still differ in several points.

(Buffon, 1770)

The journal, *Voyage et Avantures de François Leguat* (1708), is full of invaluable information. Although Leguat was not a trained naturalist, he knew how to observe, and his statements have been confirmed by research undertaken more recently:

It is very fortunate for naturalists that François Leguat had to live for two years marooned in a desert island of the Indian Ocean, and that he had published the observations he had made during that prolonged period.

(Milne-Edwards, 1896)

Leguat has described [the Solitaire] *superbly, to the minutest detail, as an observer who, though not a naturalist in the strict sense of the word, was at once attracted by the charm and dignity of a unique creature as yet fearless of man. For months he observed it. His fascinating description is remarkable for its precision, elegance and simplicity. I have never come across anything of comparable excellence in travellers' descriptions of Mascarene extinct avifauna. It has the unmistakable stamp of authentic originality.*

(North-Coombes, 1979)

Courtesy of the North-Coombes family.

Alfred North-Coombes (1907-1998)
Alfred was born in Mauritius, and studied classics at the Royal College of Curepipe. He later studied at the College of Agriculture and won a scholarship for further studies in 1928, obtaining a BSc in agriculture at the University of Reading, England, in 1931. He returned to Mauritius to a lectureship at the College of Agriculture, but then moved to the Department of Agriculture, of which he later became director. Alfred was made an OBE in 1956 for services to agriculture. He immigrated to Australia in 1963 where he gained a doctorate for his thesis, *The Vindication of François Leguat*. When he returned to Mauritius in 1988 he was given an honorary professorship of the University of Mauritius and made an honorary member of the Société de l'Histoire de l'Île Maurice.

Leguat admitted that he fell under the charm of what became his favourite bird:

Of all the birds in the Island the most remarkable is that which goes by the name of Solitary, because it is very seldom seen in Company, though there are an abundance of them.

Left: Drawing of the Solitaire by François Leguat, from his *Voyage et Avantures* (1708).

He firstly described the differences between the male and female Solitaires:

The colour of the males' plumage is ordinarily grey and brown. The feet as well as the beak are like a turkey, but the beak is a little more hooked. They hardly have any tail and their hind part, covered with feathers, is round like a horse's rump. They are taller than turkeys, and their neck is straight, and a little longer than that of a turkey when it raises its head; the eye is black and the head is without comb or crest. They never fly and their wings are too small to sustain the weight of their bodies; they serve only to beat themselves, and flutter when they call one another. From March to September they are extremely fat … one finds males which weigh as much as 45 pounds [20 kg].

Of the females, some are fair like the colour of blonde[4] hair and some are brown. They have a kind of headband, like the headband of widows, high upon their beak which is of tan-colour. The feathers are all carefully plumed and those on their thighs are rounded at the end like shells and are very thick there.

They have two elevations on the crop, of which the feathers are whiter than the rest, and which resemble, very marvellously, the beautiful bosom of a woman. They walk with such stately form and good grace that one cannot help admiring and loving them.

Solitaire by Eric Kwet-On (1999).

It is interesting to note that in his translation of Leguat's work, Georges-Louis LeClerc, Comte de Buffon (the bashful Victorian scientist and writer), translates the French word *sein* as 'neck' instead of 'bosom', which Leguat was describing as almost white in colour.

Here, Buffon (1770) gives us a condensation of the differences in colour, as seen by Leguat:

The plumage of these is commonly mixed with grey and brown; but in the females, sometimes brown, sometimes a light yellow, predominates. Carré says that the colour of the plumage of these birds is glossy, bordering on yellow; he adds, that it is exceedingly beautiful.

---

4   The blonde birds were no doubt semi-albinos, or what ornithologists would call cinnamons, which are usually hen birds in the wild.

Like the Dodo, it seems that the young Solitaire squabs took several years to attain their complete plumage. This pattern can be observed all along the British coast where the colour of adult seagulls is grey and white, whereas the young tend to be a spotted-brown colour.

Leguat describes the feathers of the females stating that *'No one feather is straggling from the other all over their bodies, they being very careful to adjust themselves, and make them all even with their beaks.'* It also seems that the feathers had a hair-like feature, similar to that of the kiwis of New Zealand.

As far as the tail was concerned there was very little of it. The female had a slight pointing of some small tail feathers, whereas the male tail feathers seemed larger and longer. The wing feathers were about five in number, being long, horizontal and held against the body.

After François Leguat left Rodrigues in 1693, there were few visitors. The only other detailed eyewitness account of the Solitaire is by Julien Tafforet, a native of Réunion, who was second mate on the ship *La Ressource*. He had been sent to Rodrigues to claim and explore the island for France in 1725. Due to bad weather, he and four other men were marooned there for nine months. During their stay Tafforet observed the bird, and later described it in the document, *Relation de l'isle Rodrigue*:

The Solitary is a large bird, which weighs about forty or fifty pounds. They have a very big head, with a sort of frontlet, as if of black velvet. Their feathers are neither feathers nor fur; they are of a light grey colour, with a little black on their backs. Strutting proudly about, either alone or in pairs, they preen their plumage or fur with their beak and keep themselves very clean. They have their toes furnished with very hard scales, and run with quickness, mostly among the rocks, where a man, however agile, can hardly catch them.

Solitaires in a forest. Composite picture by Bob Latimer.

They have a very short beak, of about an inch in length, which is sharp. They nevertheless do not attempt to hurt anyone, except when they find someone before them, and when hardly pressed try to bite them. They have a small stump of a wing which has a sort of bullet at its extremity, and serves as a defence. They do not fly at all, having no feathers to their wings but they flap them and make a great noise with their wings when angry and the noise is something like thunder in the distance.

They only lay, as I am led to suppose, but once in the year, and only one egg. Not that I have seen their eggs, for I have not been able to discover where they lay. But I have never seen but one little one alone with them, and if anyone tried to approach it, they would bite him severely. These birds live on seeds and leaves of trees, which they pick up on the ground. I have eaten them; they are tolerably well tasted.

Tafforet's description of the *'bullet'* on the wing confirms what Leguat observed as a *'little round mass under the feathers as big as a musket ball'*. Frederic Lucas (1893) points out that both these observations have been confirmed by evidence in fossil remains.

A very brief account of the Solitaire was given in 1735 by Gennes, when men from his ship went ashore on Rodrigues to replenish food supplies and returned with some game:

Our men told of having seen goats and a large quantity of birds of different kinds; they brought, amongst others, two which were bigger by a third than the largest turkey; they appeared, nevertheless to be still very young, still having down on the neck and head; their wingtips were but sparsely feathered, without any proper tail. Three sailors told me of having seen two others, of the same species, as big as the biggest ostrich. The young ones that were brought had the head made more or less like the latter animal, but their feet were similar to those of turkeys, instead of the ostrich which is forked and cloven in the shape of a

hind's foot. These two birds, when skinned, had an inch of fat on the body. One was made into a pie, which turned out to be so tough that it was uneatable.

(Cheke, 1987)

The mention of fat on the body ties in with Leguat's comments and a recent study by Staub (2000) – discussed in the section on food, in *chapter 6*. However, Leguat and Gennes differ regarding the taste of the Solitaire, which Leguat found to be 'excellent'.

It may be worth considering what the Solitaire looked like in comparison with the more widely known Mauritian Dodo. As would be expected, the two birds took on their own individual differences, having journeyed their separate ways so long ago. These differences were demonstrated by the general proportions of the body, the skull formation, and the sexual dimorphism.

Strickland and Melville (1848) compared the few bones of the Solitaire and the Dodo and came to the conclusion *'that they are not specifically identical. The tarso-metatarsal from Rodriguez is about an inch longer than that of the Dodo, and the proportions of the other bones indicate a more erect and longer legged bird, precisely as the description and figure of the Solitaire given by Leguat would lead us to expect. On the other hand* [the bones show that] *the species to which they belong is unquestionably very nearly allied to, though not identical with, the Dodo.'*

Hachisuka (1953) mentions that:

> ... *the bones of the wing in Pezophaps are more massive and smoother than in the true Dodo. All that can be said with certainty of the appearance of the Solitaire seems to be that it had neither the awkwardness nor the large beak of the Dodo, while the head was flat rather than raised at the top. The general colour of the males was brownish grey and darker on the back. The eyes were black, and according to the writer of the* Relation de l'isle Rodrigue *the frontal band was like black velvet. It cannot be determined with certainty whether the tan-colour of Leguat refers to the band or to the bill.*

Bradley Livezy (1993) informs us that *'of the two, the Solitaire was the heavier bird – the average weight of males is estimated at between 20.9 and 27.8 kg, compared with 15.9 and 21.2 kg for the Dodo. One striking feature of both species is the exaggerated sexual dimorphism; females were only two-thirds as heavy as males, and had much shorter bills.'*

Hume (2003) maintains that sexual dimorphism in the Solitaire is more extreme than in any other carinated (keel-bearing) bird.

Storer (2005) reflects that *'almost everyone who has had the opportunity to examine skeletons of these two birds has been strongly impressed with the differences between them.'*

See *Appendix D* for a summary of the main physical and skeletal differences between the Rodrigues Solitaire and the Mauritian Dodo.

Solitaire painted by Vaco Baissac, Vaco Art Gallery, Grand Bay, Mauritius.

6 Some peculiarities

Flightlessness

As the larger ground-feeding birds seldom take flight except to escape danger, I believe that the nearly wingless condition of several birds, which now inhabit or have lately inhabited several oceanic islands, tenanted by no beast of prey, has been caused by disuse. (Darwin, 1859)

There is no doubt that the Solitaire was a flying bird when it arrived at the remote island of Rodrigues. The ability to fly is a powerful skill, which allows birds to:

- Reach foods inaccessible to other animals
- Escape predators
- Build their nests in safe places away from other animals
- Perch and sleep high up above ground level
- Travel long distances for food and migration.

So why would the Solitaire give up flight? Its new habitat of Rodrigues offered an abundance of food on the forest floor or near to the ground; there were no predators from which to escape; the ground was safe for building nests; and it did not have to travel very far for any of its needs.

Grzimek (1968) suggests that *'it was the life on the islands which brought flightlessness. Such a great modification is not unique in the birds, for it is also found in ostriches, cassowaries, the nandu, various rails, and other species. Body parts and organs which have become extraneous are generally modified or degenerate.'*

Watercolour by George Richmond, 1840.

Charles (Robert) Darwin (1809-1882)
Charles was an English natural historian and geologist and a proponent of the theory of evolution by natural selection. He was an unpaid naturalist on HMS *Beagle* for her voyage around the southern hemisphere (1831-6), when he collected the material that became the basis for his ideas on natural selection. In 1859, he published *On the Origin of Species* and in 1871, *The Descent of Man.* He visited Mauritius in 1836 but does not mention the giant tortoises or the Dodo, which were already extinct.

Left: Nest with egg, male (right), female (left), from *The Dodo and Kindred Birds* by Hachisuka (1953). Courtesy of Errol Fuller and IPC

Birds make flight look easy, but it takes a considerable amount of effort for heavier birds to even lift themselves off the ground. They also need a long, uninterrupted take-off area, which was not easily found in the deep forests and on the hilly ground of Rodrigues. Over thousands of years, the increase in weight made it very difficult for the Solitaire to take off, and it didn't need to anyway.

As Attenborough (1998) remarked, '*birds have repeatedly abandoned flight during their history. Flying is very expensive in terms of energy and birds do not travel by air if it is safe for them to do so by land.*'

This energy conservation allows the bird to gain weight and one of the advantages of this, as highlighted by Cooper (in Angier, 2002), is that '*in an island situation, the birds that put on the most mass, and eat the most, can dominate in terms of mating and territory.*'

Birds can give up flight very quickly in evolutionary terms; the process may take only 10,000 years. During this time the

Order 6.Genus 5, Dodo.   *Order 6.Genus 50, Ostrich.*   *Order 6.Genus 59, Cassowary.*

Other flightless birds, detail from an art print (c. 1795). Courtesy of IPC Ltd.

Solitaire would have gradually adapted itself to life on the ground, where the only conflict would come from other Solitaires fighting over their territory at nesting time.

Stones

They have a gizzard larger than a fist, and what is surprising is that there is found in it a stone of the size of a hen's egg, of oval shape, a little flattened, although this animal cannot swallow anything larger than a cherry stone. (Tafforet, 1726)[5]

Birds do not have teeth. This is partly due to the fact that teeth are heavy, and flying birds do everything in their power to shed weight, and partly to do with aerodynamics, as a bird will nose-dive if it has a mouthful of teeth. So, birds grind up food in a gizzard, which is more centrally placed in the body, and they swallow small pieces of grit to help in this grinding process. Birds that do not fly can afford to have larger stones in their gizzards, since weight does not create a problem, as Leguat observed in the Solitaire:

We find in the gizzards of both male and female, a brown stone, of the size of a chicken's egg: it is a bit rough and flat on one side, and round on the other, being rather heavy and hard. We believe that this stone was there when they were hatched, for even the very young ones have it. They only seem to have one of them, and it is a mystery as the passage from the crop to the gizzard is so narrow, that an object of half the bigness

[5] The writer may not have been correct in his statement, as many pigeons have very flexible lower mandibles, enabling them to swallow quite large fruits.

couldn't pass. It served to sharpen our knives better than any other stone.

D'Heguerty (1754), a governor of Réunion, wrote that when dissecting a Solitaire *'one ordinarily finds stones which are considered valuable and which are useful in medicine.'*

Stones similar to those found in the gizzard of the Solitaire.

Caldwell (1875) comments on the accuracy of Leguat's observations in his report on lifting bones from the floor of one of the caverns in Rodrigues:

I got, both with the mounted bird and the male bird, the stones mentioned by Leguat as existing in the gizzard. In each case they were found on lifting the sternum and in the middle of the ribs. They are basaltic pebbles with rough angles and surfaces, and no stone of similar kind is to be found within two miles of the caverns. I got four in all, but only two of which I could identify the birds they belonged to.

Sir Hamon L'Estrange was passing an exhibition in London in 1638 when he saw a Mauritian Dodo in captivity and was tempted to investigate. He later recorded that he saw *'a heap of large pebble stones … some as big as nutmegs and the keeper told us that she eats them.'* The stone found in the Solitaire seems to have been of a similar size. It is not certain whether there was one stone or two, but the stone, or stones, were foreign to the bird and were not formed inside, although it was reported that young birds also had a small stone. Birds that do not fly very much, like chickens, turkeys and geese, have quite large gizzards, filled with small stones and grit. Birds that do not fly at all, like the Ostrich, pick up and swallow quite large stones, and the extinct moas of New Zealand were known to swallow stones that were 10 cm across. There has been no 'stone-making' mechanism found in any living bird, therefore it is thought that all stones found in birds come from an external source.

Nests and eggs

When these birds build their nests, they choose a clean place, gather together some Palm-leaves for that purpose, and heap them up a foot and a half high from the ground, on which they sit. (Leguat, 1708)

Leguat described the nesting habits of the bird:

They only lay one egg, which is much bigger than that of a Goose. The male and female both cover it in turns, and the young is not hatched for seven weeks. All the while they are sitting upon it, or are bringing up their little one, which is not able to provide for itself for several months, they will not suffer any other bird of their species to come within two hundred yards round of the place. But what is very singular is the males will never drive away the females, only when he perceives one he makes a noise with his wings to call the female, and she drives the unwelcome stranger away, not leaving till 'tis without her bounds. The female do's the same as the males, whom she leaves to the male, and he drives them away. The combat between them on this occasion lasts sometimes pretty long, because the stranger only turns about, and does not fly directly from the nest. However, the parents do not rest until they have driven the intruder from their limits.

Tafforet (1726) also mentions that the Solitaire lays one egg a year. This is based on the fact that he only saw one squab accompanying the adult, as he admitted that he had never found an actual nest.

A single white egg was laid on a ground nest.

Leguat observed that the egg was larger than that of a goose (probably referring to the domestic White Goose, which has an egg length of about 10 cm). A Dodo egg was described as being *'the size of a penny bun … or the egg of a White Pelican'* (Cauche, 1651). The size of a penny bun is difficult to determine, but the nearest living relative to the White Pelican is the Pink-backed Pelican whose egg size is 6 x 9.2 cm. The fact that only one egg was laid and reared at any one time helps us to understand why the Solitaire disappeared so quickly.

The Solitaire may have reared its young during the months of May to August, this is based on the premise that it would have bred during the palm fruiting season, when conditions were better. From September, the palm fruit would have stopped dropping and the Solitaire would have started to moult. The incubation period would have been about seven weeks, and the young would have taken about nine months to mature (Staub, 2000).

Marriage

We have often remarked, that some days after the young one leaves the nest, a company of thirty or forty brings another young one to it; and the new fledged bird with its father and mother joining with the band, march to some place nearby. We frequently followed them, and found that afterwards the old ones went each their way alone, or in couples, and left the two young ones together, which we called a Marriage. (Leguat, 1708)

A 'marriage'. Composite picture by Bob Latimer, from an original wall painting by Julian Hume.

Some critics of Leguat's work pounce upon the statement that there were some *'thirty or forty'* birds together in the mating ceremony, and remind us that the bird is called a Solitaire for its solitary existence. Also, his observation that a 'marriage ceremony' took place is perhaps his way of seeing what he wanted to see with regard to the behaviour of the birds he admired so much.

He also reported that the Solitaires were monogamous and stayed together for life:

After these birds have raised their young one, and left it to its self, they are always together, which the other birds are not, and though they happen to mingle with other birds of the same species, these two companions never disunite. This particularity has something in it which looks a little fabulous; nevertheless, what I say is sincere truth, and what I have more than once observed with care and pleasure.

Usually birds that show sexual differences in size are polygamous, and not monogamous as described by Leguat. However, we have seen the general accuracy of Leguat's observations and so there is no reason to doubt this one.

Defence

The young bird was apparently dependent on food provided by its parents for several months, and the defence of an adequate food supply would have ensured that they successfully reared their fledgling.

(Halliday, 1978)

The territory of the Solitaire seemed to be something up to 200 m, with the nest in the centre. Both members of the pair defended it, by chasing off intruders. The birds also used a dramatic display of pirouetting and vibrating their wings violently to produce a sound that could be heard all over their territory. Leguat tells us that the wings *'serve only to beat themselves and flutter when they call one another. They will whirl about for twenty or thirty times together on the same side, during the space of four or five minutes. The motion of their wing makes them a noise very like that of a rattle; and one may hear it two hundred paces off.'*

If this act didn't frighten off other birds, then they used their wings to beat off their opponents. The many Solitaire bones found in the caves have now proved that Leguat was correct when he wrote about a ball-like lump at the end of the wing, which was used for fighting:

The bone of the wing increases in size toward the end and forms a small round mass under the feathers as big as a musket ball. That and its beak are the chief defence of this bird. It is very hard to catch it in the

Solitaires guarding their territory. Acrylic on paper by Julian Hume. Courtesy of IPC Ltd.

Woods, but easy in open places, because we run faster than they, and sometimes we approach them without too much trouble.

(Leguat, 1708)

Many animals run away as a form of defensive behaviour, and the Solitaire seemed able to do this in the forest, but not on open ground:

It is difficult to ensnare in the woods, where it can elude the sportsman by cunning and dexterity in concealing itself, but as it does not run fast, it is easily caught in the plains and open fields; when overtaken, it utters not a complaint, but wastes its grief in tears, and obstinately refuses every kind of food. (Buffon, 1770)

Lucas (1893), who was interested in the fighting methods used by the Solitaire, refers to Leguat's description of the musket ball and makes his own observation:

[Leguat] *very aptly describes the swollen bone at the base of the metacarpus, and this, swung by the short, stout little wing, must have been capable of hitting a pretty hard blow, even if, as is probable, it was surrounded by thick, callous skin. The outer end of the forearm (radius) is also rough and swollen, and it looks very much as if this enlargement of the bone had originally been brought about by the Solitaire's combative habits, the wrist joint having been banged and bruised until that diseased outgrowth known as exostosis took place, and became a constant character of the bird.*

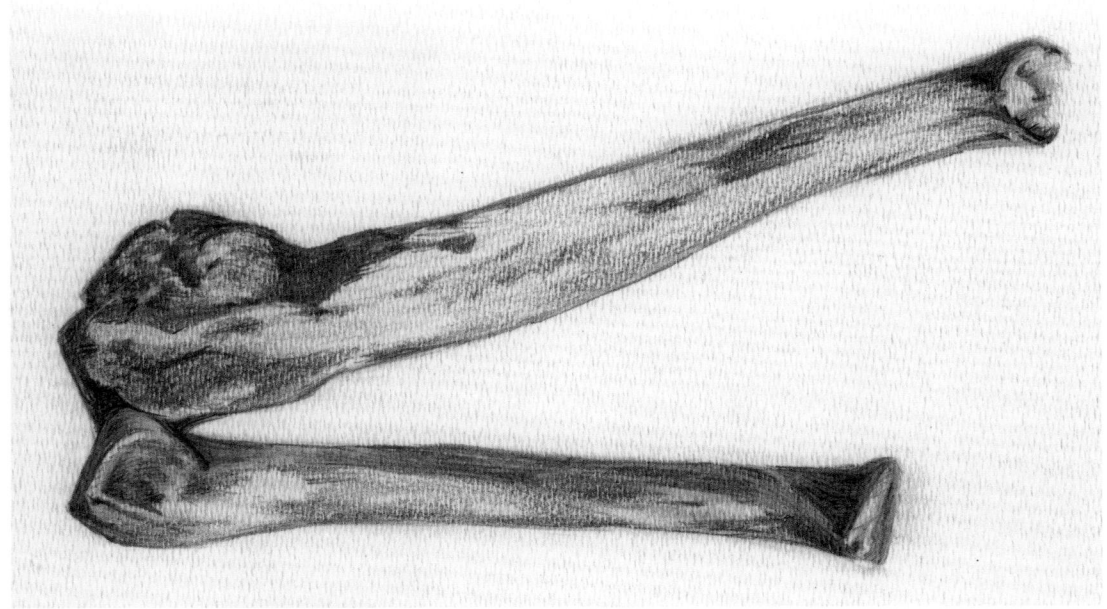

Wing bones showing 'musket ball' growth. Pencil drawing by Kathleen Latimer.

Richard Owen (1878) adds his own comment about this weapon:

> This hard, irregular, prominent mass, which holds the place of the spine in the Spur-winged Goose, may be compared to a 'knuckle-duster'; with it the combative sex delivered his blows, in the hard and well-contested fights to which Leguat testifies.

An interesting observation made by Strickland and Melville (1848) was the unusually large keel protruding from the sternum, which serves as an attachment and insertion area for the well-developed pectoral muscles needed for flight. Leguat testified that the bird did not fly but that it used its wings for self-defence, inflicting considerable blows with them, which would have needed a sternal keel and strong pectoral muscles for successful operation.

Food

> Early travellers seemed to be more interested in satisfying their own hunger pangs than observing what the local fauna ate.

Leguat says little about the feeding habits of the Solitaire, but the plantain dates and latan fruit probably would have been favourite foods for the birds as well as the settlers:

> The Plantain is a sort of a palm-tree … the dates of which are bigger than those of the actual palm-tree. Having abundance of better things to feed on, fish, flesh and fruits etc, we left the dates for the Turtle Doves and other birds, particularly the Solitaire.

The Solitaire probably ate the dark green acorn fruit of the Rodrigues endemic ebony. The Mascarenes have fourteen endemic

Fruits on *Pittosporum balfourii* (endemic), which the Solitaire may have eaten.

species of the genus *Diospyryos*: twelve occur in Mauritius and one on each sister island. When Leguat was on Rodrigues there were over forty endemic plant species, including pandanus and three endemic palms. He tells us that the valleys were full of palms, plantains, ebonies and many other trees, which would have dropped their fruit to the forest floor.

Plants fruit profusely in Rodrigues from June to August, and these tended to fatten the Solitaires, inducing them to lay their eggs in that season. From October to December a drought usually occurs, preceding the January to May rainy spell that prompts flowering.

Leguat comments that *'from March to September (winter months), Solitaires are extraordinarily fat and their taste is excellent, especially the young.'*

Tafforet (1726) describes how the Solitaires survived during the fruitless period in December, living on seeds and leaves of trees, picked up off the ground.

A study by France Staub (2000) looked at the food available for the Solitaire during the winter and summer seasons. He suggested that during the summer the Solitaire favoured the areas where the forest abounded with latanier trees, and where their feasting would enable them to build up a small surplus of fat. According to Staub, analysis has shown that these palm, latanier and pandanus fruits are rich in oil, which thus explains *'the secret of quick fattening in the … Solitaire'*.

The Pink Pigeon in Mauritius similarly builds up a surplus of fat, but the extra size and weight cannot be detected by the casual onlooker. Many scientists disagree with

Fruits on *Ficus rubra*, a native plant of Rodrigues.

Staub's theory, as this would have meant the storage of improbable amounts of fat, and the bird would have had to starve for five months to slim down to its leaner self. It is thought that this would have been almost impossible for any large bird, with such a slow metabolic rate, living on a subtropical island.

Not for export

It seems strange that not one Solitaire appears to have been captured and taken by ship to another part of the world.

The adventurous Dodo of Mauritius found its way to various corners of the earth. There are records of nine being shipped to Holland; one was painted live in Prague (capital of the Habsburg Empire); one was exhibited in London; another was sent to Japan; and two live Dodos ended up in a menagerie belonging to the Mogul Emperor Jahangir of Surat (West India). However, it seems that the Solitaire was not as intrepid as its travelling cousin, as Leguat noted that:

Although these birds will sometimes come up near to one when we don't run after them; they will never grow tame. As soon as they are caught they shed tears without actually crying, and refuse all food and water until they die.

From the Royal Society of Arts and Sciences

France Staub (1920-2005)

France was a Mauritian dentist who had a lifelong interest in ornithology, orchids, history and art. He was brought up in an area called Staub on the family sugar estate at Baie du Cap. This area was cut off from the rest of Mauritius and travel was usually by coastal boat. He attended St Joseph's College, and in 1944 he graduated in science from the Agricultural College (later to become the University of Mauritius), before attending Guy's Hospital Medical and Dental School in London, where he qualified as a dental surgeon in 1951. While at agricultural college he came under the influence of Jean Vinson, who was the only zoologist on the island. Vinson used to take his students out on field trips, and when France returned to Mauritius from England Vinson invited him to join the local natural history group, the Royal Society of Arts and Sciences (est. 1829), in which he remained active for the rest of his life, joining its council in 1964, and serving as its president several times.

In the 1960s, he often accompanied Vinson and other local naturalists on filming and bird-ringing trips to Round Island, and on expeditions to the remote islands of St Brandon (Cargados Carajos), Rodrigues, Agalega and the French island of Tromelin. In 1973, he published the first of a series of popular bird books on the Mascarenes, with revised and expanded versions appearing in 1976 and 1993 – the last including the entire vertebrate fauna and more. His last ornithological publications were in the 1990s, when he explored the importance of seabirds and guano in the history of sugar cultivation, the co-evolution of endemic birds with nectar-flowers and fruit, and speculated on the life cycle and ecology of the Dodo.

It seems that the Solitaire was extremely traumatised by captivity, which the following observation by D'Heguerty (1754), confirms:

> One also finds birds of different species, which can be caught by running after them, *and among them the Solitaires, which scarcely have tails or wings; this bird, as big as a Swan, has a sad face; in captivity one sees him always in the same line, no matter how much room he has, and returning the same way, without variation.*

The Solitaire, unlike the Dodo, was not a good traveller. From *The Oldest Dodo in the World* by Jane Lagesse, illustrated by Stina Spangenberg Becherel.

7 Digging for bones

… perhaps no species has had its osteology examined on so great a scale as the Solitaire. (Newton and Gadow, 1896)

After Julien Tafforet left Rodrigues in 1726, other eyewitnesses paid little heed to the Solitaires, which had gradually withdrawn into the forests, leaving visitors to marvel at the thousands of tortoises on the island.

By the time that Pingré visited Rodrigues in 1761, the Solitaire had almost disappeared altogether, and all that was left to interest visitors was to search for bones to provide proof that the Solitaire had actually lived at all. The search for these remains focused on the vast limestone caverns discovered on the island, where many bones were found buried in the dry soil, protected from the corrosive nature of rain.

Strangely, neither Leguat, nor Tafforet, nor Pingré made any mention of the caves, although all three men had explored the island thoroughly. Obviously the limestone plain, where the caves are found, kept their existence a secret.

Much later, Balfour (1879) was able to tell us of the many caves found on this plain:

The caves from which the bones of the Solitaire and other extinct birds have been obtained occur in this limestone plain … The whole plain is riddled with caves, and in walking over it one constantly passes small apertures and fissures, evidently 'blow-holes' of some subterranean cavern.

The first search for bones was conducted as early as 1786 by Captain Labistour, who found what were thought to be Dodo bones encrusted with stalagmite residue in caves at Plaine Corail. These bones did not get passed on until 1830, when they were given to Julien Desjardins, secretary to the Natural History Society of Mauritius. He in turn passed them on to the naturalist Georges Cuvier, who wrongly proclaimed that they were Mauritian Dodo bones (Cuvier, 1830).

Left: Solitaire bones left *in situ* on the floor of a small cave at the François Leguat Giant Tortoise and Cave Reserve, Plaine Corail. Photograph by Bob Latimer and Aurèle André.
Right and overleaf: Reconstructed Solitaire skeleton and skull at the Natural History Museum (Mauritius). Photographs by Bob Latimer. Courtesy of IPC Ltd.

This find aroused interest in Mauritius where Charles Telfair of the Society requested for some more digs to take place in the caves. In 1831, a resident landowner called Honoré Eudes dug up some bones in a large cavern and sent them to Telfair, who presented them to the Zoological Society of London.

One of these bones was given to Hugh Strickland, who placed it in the University Museum of Zoology, Cambridge, and others were sent to the Andersonian Museum, Glasgow. The bones were examined and it was confirmed that they belonged to the Solitaire. Finally, the eyewitness descriptions were confirmed by this new skeletal evidence.

It is interesting to note that the bones sent by Telfair to the Zoological Society seem to have disappeared soon after this date, when Hugh Strickland requested the curator of the Society to show him the bones. After searching among the many treasures presented to the Society, the original box given by Telfair was found, but the bones of the flightless Solitaire had flown away!

At that time there were very few bones for Strickland and Melville to examine from other sources. By 1852, there were only eighteen bones in Europe: five in Paris, six in Glasgow, five in the possession of the Zoological Society (since transferred to the Natural History Museum, London) and two held by Strickland himself. The bones from the Paris Museum were so encrusted with stalagmite material that it prevented a proper examination of the bone surface, and made it difficult to describe their original form and structure. It was deduced, from the uniformity of the appearance and thickness of the encrustation, that all the bones collected in Paris had come from the same locality and had been exposed to dripping water containing carbonate of lime. The bones from the Glasgow collection were in better condition and not coated with lime.

Solitaire bones can be seen at a number of museums and institutions throughout the world, a list of some of these is given in *Appendix E*.

From the two collections, it was established that the bones came from the same species of bird and that, as the samples came from Rodrigues, it must have been the Solitaire, originally described by Leguat. Due to the great difference in the size of the specimens, Strickland initially proposed that the bones belonged to two species, the second he named *Pezophaps minor*. Later, when more bones were available, the variety in size was attributed to sexual differences.

Dr A. G. Melville, who wrote the osteological part of *The Dodo and its Kindred* (1848), concluded that:

> *We have now ascertained that the cranium of the Solitaire resembles that of the Dodo in numerous important points, differing in such respects only as would justify us in regarding these birds as specifically distinct … The marked dissimilarity in external form between the Dodo and the Solitaire and the position of the caruncular ridge in the latter, together with the shorter beak, fully justify the establishment of another genus Pezophaps in the Didinae, to include the lost form. That the Dodo and Solitaire belong to the same extinct sub-family of the Columbidae, characterized by the peculiar structure of the cranium and rudimentary wings, no one will, we trust doubt, who has carefully and impartially examined the evidence.*

In 1864, Edward Newton, who was working in Mauritius, sailed to Rodrigues on the HMS *Rapid* to study the birds and to explore the caves. It took some effort to organise an excursion to the caves at that time, as they sailed by boat from Port Mathurin, and often had to wait for suitable tides. On this particular occasion the party set off at one o'clock in the morning and Newton (1865) reported that at *'about six o'clock we landed, and at once walked up to the first cave, about a quarter of a mile inland'*.

They found only a few crumbling tortoise bones near the entrance of this first cave, but they sailed round the coast for about 5 km back towards Port Mathurin, and walked inland for 3 km to a second cave, where they picked up a couple of bones on the surface near the entrance. Much to Newton's disappointment, they had to rush to catch the tide:

> *The next morning, to my disgust, it was decided to return to Mauritius immediately … thus ended my visit to Rodrigues, where I could have well managed to spend a month to my own advantage.*

The frustrated Edward Newton urged a magistrate on the island, by the name of

Courtesy of Errol Fuller, *Extinct Birds* (2001).

Sir Edward Newton (1832-1897)
Edward was the younger brother of Alfred, the zoologist at Cambridge. He entered the colonial service in 1829 and was posted to Mauritius, where he became assistant colonial secretary (1868), and then colonial secretary until 1877. He helped George Clark search for bones, and was the founder of the *Ibis Ornithological Review*.

George Jenner, to explore the caves for more bones, and by 1865 Jenner had sent Newton a box of bones containing the remains of tortoises and large birds. Jenner continued to dig, but because he found it difficult to recruit superstitious Rodriguans, who were unwilling to go into the dark caves, he was sent four Indian labourers from Mauritius. These men, working under the supervision of a Sergeant Morris, returned to Mauritius after two months with nearly 2000 bones or bone fragments of the Solitaire and other animals. These were studied by the Newton brothers, who wrote a paper on the Solitaire and read it to the Royal Society in 1868.

George Jenner
George served with the 4th Regiment from 1848 until 1858, and fought in the Crimean War. He was appointed police inspector of Mauritius in 1858 and was transferred to Rodrigues four years later, where he became a magistrate. He started the first school on the island, and implemented the appointment of a Roman Catholic priest. He served in Rodrigues for nearly ten years, and in 1871 left to become the sanitary warden of Port Louis in Mauritius. One of the main streets in Rodrigues is now called Jenner Street.

The next bone explorer arrived in 1874, when the second transit of Venus expedition went to Rodrigues with a team of scientists to study various aspects of its natural history. The man selected for researching the extinct fauna was the Revd Henry Slater. With the help of Edward Newton in Mauritius, he was given ten men, including a cook, to help dig in the caves, which he described:

Courtesy of Errol Fuller, *Extinct Birds* (2001).

Alfred Newton (1829-1907)
Alfred was English, but was born in Geneva, Switzerland. In 1854, he was elected travelling fellow of Magdalene College, Cambridge, and in 1866 became professor of zoology and comparative anatomy. He was awarded a Royal Medal by the Royal Society in 1900, for his services to ornithology and zoogeography. He made a study of the Dodo, aided by his brother Edward, who was stationed in Mauritius. In 1872, he was the first person to describe Newton's Parakeet, which lived on Rodrigues.

The cave tract in Rodrigues is situated about the S.W. side of the island and is of a very curious nature. We find there 10 to 12 patches of limestone scattered upon the basalt which forms the island. On examination, these patches are found to consist of marine coral upheaved with the basalt.

It is in these coralline limestone patches solely that the caves are situated. The depth of the bone-earth is very variable; in some caves we find it with a depth of from 6 inches to 3 feet; in others it varies from 4 to 9 feet in depth. Below 2 feet I never found many bones. (Slater, 1879)

For convenience, Slater camped near the caves and each evening he went out to shoot game, in the process of which he discovered many new caves:

These formed a branch of a cavern already dug, but as entrance to them was extremely difficult, they had hitherto escaped notice. In these we found a large quantity of Solitaire bones and the almost perfect skeletons of a male and female … very few tortoise bones were intermixed, and I found it almost an invariable rule, that where Solitaire bones were found in large numbers and apparently occupying the spots where they died, there were few or no tortoises among them … On 23rd October … I was waiting for my headman near a curious ravine which we called 'the Gorge', intending to make a survey of it with a view to new caves, when I noticed a small dark hole behind a huge block of coralline limestone.

Henry H. Slater
The Revd Henry Slater was one of three naturalists attached to the expedition to observe the passing of the planet Venus on Rodrigues in 1874. The Royal Society, on behalf of the British Government, asked Slater to explore the caverns with the special task of studying the extinct fauna, while Isaac Balfour was the botanist and geologist, and George Gulliver was requested to investigate the fauna generally.

This small hole led to a large cave, which had been hidden by fallen rock. Here he found many Solitaire bones, including a skeleton uncovered in a small crevice where the bird had been trapped. Two weeks later another new cave was discovered, also rich in bones, but after this no more caves of any value were found. One big disappointment for Slater was that he was unable to find any of the gizzard stones reported by Leguat.

Shortly after Slater's discoveries, William Caldwell explored the caves in a three-month tour of Rodrigues in 1875. He had been summoned to the island to look into charges of fraud made against a magistrate by the name of Henry Reid Bell, who turned out to be one of the biggest rogues in the history of Rodrigues.

'The Gorge'. François Leguat Giant Tortoise and Cave Reserve.

Caldwell (1875) was able to find four gizzard stones, which Slater had missed, but at first he only found two Solitaire bones as Slater had done his work well. Then, as luck would have it, on the way back from a search, Sergeant Morris squeezed into a small hole to reach for several semi-fossil shells, when he saw a complete tibia:

Of course we all entered, and found the hole to be the entrance of a small but very well formed cavern of three stories formed like steps, none of the chambers being more than ten feet square, and close alongside one of the large caverns in which a mass of bones had already been found. How it had been overlooked, I am at a loss to conjecture.

From the scattered bones found, Caldwell was able to construct an almost perfect skeleton of a Solitaire.

The Governor of Mauritius, Frederick Napier-Broome, visited Rodrigues in 1881 and managed to obtain a collection of bones from William Vandorous, the pilot of Rodrigues. He authorised the magistrate to spend money on further research, which resulted in more bones being found and given to the Royal Society of Arts and Sciences of Mauritius.

As a result of the study performed on the osteological remains found in the caverns of Rodrigues, it was confirmed that the original descriptions of the Solitaire given by Leguat were very accurate. His eyewitness accounts and drawings give us the most accurate evidence from the life of any of the extinct non-flying birds of the Mascarene Islands. It is perhaps fair to say that the bones of the Solitaire have been examined more than those of any other bird, including the Dodo.

The excitement of discovering Solitaire bones in Rodrigues was partly dampened by the fact that in Mauritius no Dodo bones had ever been found. By 1859, a schoolteacher called George Clark had written an article in the *Mauritius Register* entitled 'A ramble round Mauritius', in which he briefly

mentions the Dodo, and clearly thought that the bones found in the caverns of Rodrigues belonged to the same species.

Hastings Museum and Art Gallery, England.
Courtesy of IPC Ltd.

George Clark (1807-1873)

George Clark spent his childhood in Wedmore, Somerset (England), where his father was rector of St Mary's Church. He married Mary Slocombe in 1832, but it is not clear whether Mary died before George travelled to Mauritius with a group of missionaries under the Mico Trustees in about 1838. He was appointed as a teacher at the diocese school in Mahébourg in 1851, at £177 per year, together with his new wife, Jane Pitt, at £48 per year. He had read about the Dodo in Strickland and Melville's book (1848), and was keen to prove that the Dodo had actually existed. He resigned in 1872 after twenty-one years of teaching at the diocesan school. George and Jane had six children, and many of their descendants can still be found in Mauritius within the Oxenham, Mackie, and De Sournay families. George died in Mahébourg in 1873 and is buried, together with Jane Pitt, in the Western Cemetery, Port Louis. There is no mention of his valuable contribution to teaching or his important discoveries at Mare aux Songes, which have contributed so much to our knowledge of the Dodo.

Clark spent most of his spare time studying natural history and had made numerous attempts at finding remains of the Dodo in and around the area where he taught in Mauritius. Around this time the first railway line was being constructed and Clark spent some time following the tracks and searching the banks of railway cuttings in the hope of finding bones.

Then, in 1865, a young civil engineer by the name of Harry Higginson was inspecting the new railway line near Mahébourg, when he saw some sugar estate workers digging for peat in a marsh near the line:

Harry Pasley Higginson (1838-1900)
Harry was the son of a clergyman and was born and bred in Thormanby, North Yorkshire (England). He trained in civil engineering, and built railways in Latvia, Mauritius, India and New Zealand. He arrived in New Zealand in 1872, where he became chief railway engineer and remained for the rest of his life. He married Florence Kebbell in 1874, producing seven children. A stained glass window in Wellington Cathedral commemorates his work, and one panel depicts a Dodo.

They were separating and placing into heaps a number of bones and various sorts among the debris. I stopped and examined them as they appeared to belong to birds and reptiles, and we had always been on the lookout for bones of the then-mythical Dodo. So I filled my pocket with the most promising ones for further examination.

Higginson knew of George Clark's interest in Dodo bones, so he took the specimens to him in Mahébourg. They both referred to Strickland and Melville's book, *The Dodo and its Kindred* (1848), and were convinced that this was the first find of Dodo bones ever.

Clark tells how he went to Mare aux Songes (near Mahébourg) and, with the cooperation of the supervisor, requested some men to feel around with their feet where the water was about a metre deep. After some searching, a Dodo tibia was found and, thus encouraged, they cut away a mass of floating weed covering the deepest part of the marsh. It was here that they found most

Latimer

The author washing bones at Mare aux Songes, Mauritius.

of the Dodo bones. The bones showed no cutting or gnawing marks and had not been burnt, so it was concluded that the birds had lived in the surrounding neighbourhood, and had died a natural death.

The marsh forms a natural depression, and animals must have enjoyed this tree-lined watering hole, which was originally filled with marine sand and debris through an existing creek. When the sea finally receded it left the sand and shells at the base of the marsh, which filled with organic material. Spring water percolated through the shell-filled sand, mixing dissolved calcium into the acidic surface water, which neutralised the peat medium to preserve the bones found there.

More recent finds of Dodo and tortoise bones (2005–7) have shown how successful the marsh has been in the preservation of material such as bones, wood, seeds, pollen and other organic material.

Not surprisingly, there were differences in the bone structure between the two Dodo-type birds, which had not shared the same habitat for millions of years. The Solitaire's neck was much longer than the Dodo's, and

Excavating bones, wood and seeds at Mare aux Songes. The Dodo Research Programme 2005.

generally the vertebrae were larger. The pelvis was more rounded than in the Dodo, and closer in shape to other pigeons, but its tail area differs from any known bird. The bones of the wing were bigger than in the Dodo, and there was a bony knob on the metacarpal, which was very unusual. The skull was different, the cranium being narrower and longer and flat on top without the forward swelling found in the Dodo, and the Solitaire's beak was smaller and straighter.

Recent research work in Mauritius and Rodrigues continues to challenge scientists to find out more about the two large birds that once enjoyed their peaceful life there. Discoveries are now studied *in situ* because the clues given by the surrounding soil, wood, seeds and other materials are just as important as the bones themselves.

The Dodo Research Programme (DRP) October 2005, June 2006 and August 2007

This is a Mauritian–European research initiative with the aim of 'reconstructing the world of the Dodo and determining the factors of its demise'. The Dodo Research Programme was put together as the result of an initial find of Dodo and giant land tortoise bones in a small section of Mare aux Songes in 2005.

On Friday 28 October 2005, a research team consisting of Kenneth Rijsdijk, Frans Bunnik and Pieter Floore, with the aid of Alan Grihault (a local expert on the Dodo) and Christian Foo Kune (general manager of Mon Trésor and Mon Desert Ltd, owner of the estate), recovered vertebrate bone fauna including *Raphus cucullatus* (Dodo), at Mare aux Songes.

Since this discovery, a non-destructive geophysical survey has been carried out and twelve core holes have been drilled to give an insight into the composition of the sedimentary layers. Excavations have uncovered remains of Dodo, tortoise and various other bones, as well as seeds from palm, ebony and Tambalacoque trees.

The importance of this find lies in the fact that the bones were found immersed in a bed of mud, which contains bones of other animals that lived side by side with the Dodo, and also includes the pollen, seeds, leaves and wood of the forests surrounding the area before man came on the scene. It has been established that the bones are 3,800 years old (+/- 300 years).

Many international researchers, including Julian Hume and Beth Shapiro, have worked on this site.

Dodo bones found in highlands cave June 2007

Fred and Debbie Stone, from the USA, were looking for cockroaches in an almost inaccessible part of a cave near Bois Chéri in the highlands of Mauritius. While searching they stumbled upon a complete skeleton of a large bird, which was later identified as a Dodo by the palaeontologist Julian Hume. This discovery confirms that Dodos lived in the higher regions of Mauritius, and should also provide excellent DNA, which has not been obtainable from previous finds. Like the caves in Rodrigues, these natural tombs are now gradually giving up their treasures in Mauritius.

Many photographs were taken prior to removing the bones to a safe place. Photograph by Stone/Hume/Griffiths.

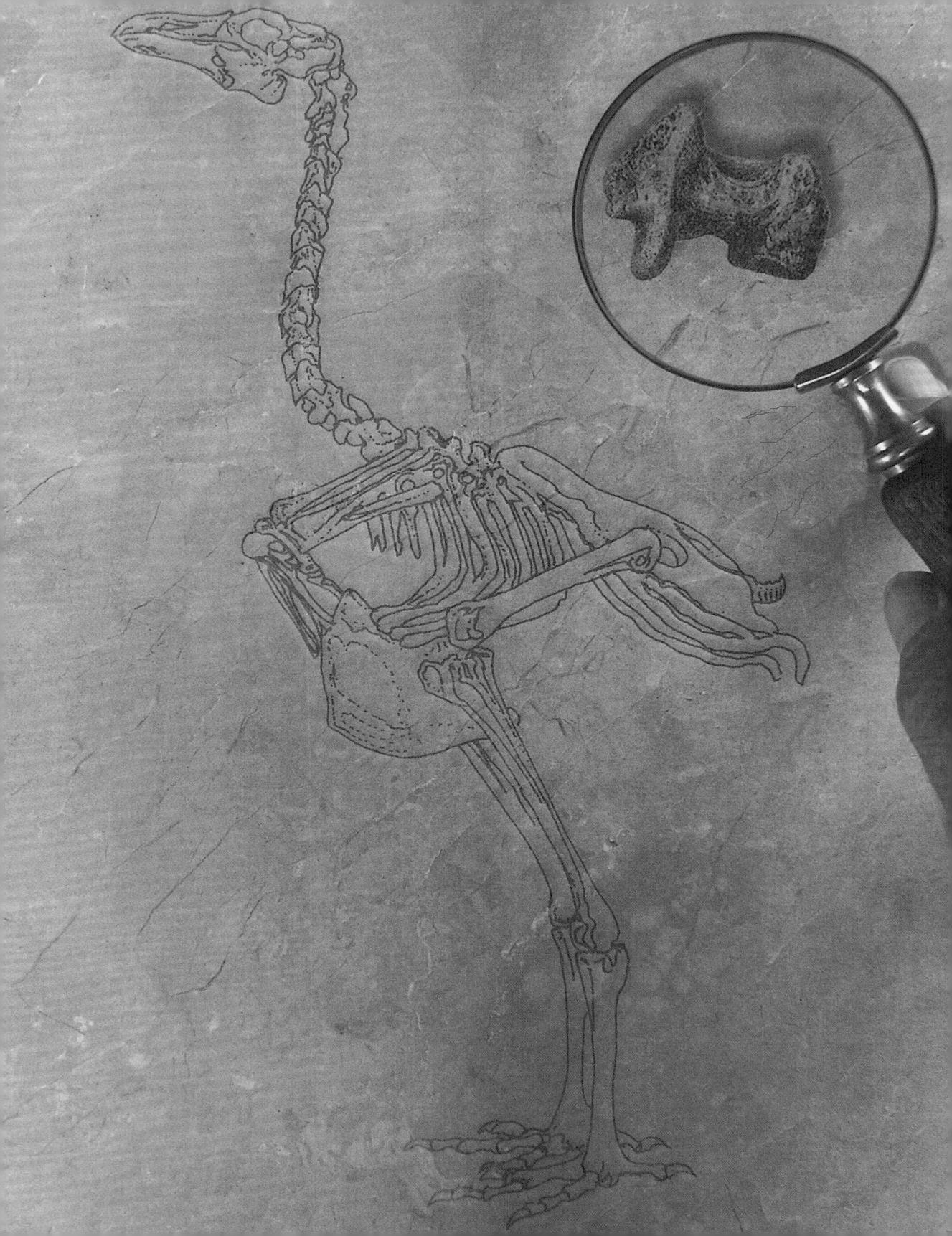

8 Post-mortem

Ever since the rather sudden demise of the Solitaire and the Dodo, scientists have tried to find a reason for these extinctions.

As far as scientists and historians can tell, it appears that the Solitaire had largely died out by the late 1760s. The birds living in the environs of Port Mathurin were ruthlessly slaughtered, together with the tortoises, and their numbers dwindled as they were forced to live and hide in the fast-disappearing forests.

By 1755, Cossigny wrote in frustration that:

… for 18 months I have been trying without success to procure a Solitaire from Rodrigues island … I have promised all one could want, in spirits or piastres, to whoever brings me at least one alive. It is claimed that cats, which have gone wild on this little island, have destroyed this species of bird that only has stumps for wings, but I am strongly inclined to believe that these cats are the men of the post who have eaten all those they have found, as they are very good …

(Cheke, 1987)

So, what were the factors that contributed to the demise of such a large and elegant bird? William Caldwell, who spent some time on Rodrigues in 1875, tells us that the Solitaire:

… lived in the midst of abundance of food, and their extinction cannot be ascribed to deficiency of nourishment, nor to human agency as the population was too sparse, and the place where their remains are now found too remote to be more than occasionally hunted; and it is well established that it is only very lately that many of the caverns in which these remains have been found have been discovered. Neither can it be granted that the bones were washed into the caverns and thus buried in the floors, though doubtless such was the case in some instances, especially in some of those explored by Mr. Slater… I can only gather therefore that these birds resorted to these caves in considerable numbers and appear to have frequented them, although this hypothesis is opposed to Leguat's statement, as he expressly mentions that the birds were not gregarious but solitary.

William James Caldwell (1820-1887)
William's whole career was spent in the public service, rising to the post of assistant-colonial secretary in Mauritius. He was sent on various missions overseas, notably to the New Hebrides (Vanuatu) and Australia, to obtain new sugar cane varieties. He was a member of the Royal Society of Arts and Sciences of Mauritius and his main interest was botany. He explored the caves of Rodrigues during a three-month tour of duty in 1875 and, from the scattered bones found there, he constructed one of the most complete Solitaire skeletons in existence.

Left: Solitaire skeleton drawn by J.-F. Sookahet. Photograph by Bob Latimer.

Caldwell ponders as to why so many Solitaire bones were found in the caves, and his questions are still unanswered today. He also responds to the suggestion that pigs may have been the reason for the ultimate destruction of the bird:

Messrs. Newton's theory of swine having destroyed them is equally, in my view, erroneous; pigs would get nothing to eat, nor water to drink, and would scarcely leave the ravines far away from this spot, where abundance of guava, raspberries, colocasia roots, and other succulent food in which they delight exist, and where they could (as at present) wallow in the muddy pools.

Looking at the barren countryside around him, Caldwell considered that '*the hypothesis that they got into the cave to avoid fires is equally untenable. Fire could not take place in this coral country, as there is no grass to propagate it and the trees are very wide apart.*'

Caldwell's supposition that there were no trees around the caves was inaccurate, since the limestone plain once supported enough vegetation to allow a fire to spread rapidly, as is suggested in the following account of a trip to the caves by Captain Labistour, in July 1786:

The valley seemed fertile; it was full of palm trees. We climbed another mountain which our guide told us was on the way to the cave … we had almost given up hope of finding it because of the harsh tracks and the trees which blocked the view in all directions. And all at once we came across the mouth of the cave, which was quite encumbered.

Caldwell (1875) hypothesises that the chief cause of the Solitaire's extinction was the occurrence of a terrible cyclone:

I can only attribute their apparently sudden and simultaneous disappearance to some terrible hurricane or other disturbing cause which led them to these places for shelter, for they are found in many places where no bird deprived of the faculty of flight, and with any instinct, would resort to, viz. withdrawn into nooks, crannies and fissures, whence they could, in many instances, scarcely get out again. How under any theory can the discovery in similar places of remains of gulls and tortoises be explained? The effect of a hurricane such as I have described, is to shake off every fruit, seed and leaf from the trees. The south-western corner of Rodriguez would be peculiarly exposed to its violence, and consequently the animals and birds living on such food would literally die of starvation.

In answer to Caldwell's claim, it seems strange for birds that had adapted themselves specifically to live on a small island like Rodrigues, which suffers frequent cyclones, to suddenly succumb to a cyclone. Even the worst cyclone possible could not endanger birds sheltering within a short distance from the entrance to the caves. There would have been no need for them to hide so far into the caverns, and they would not have found it necessary to hide in narrow crevices and holes. They surely were not escaping wind or rain in doing this, but something that would silently creep into every nook and cranny deep within the caves – something like gas, fumes or smoke.

North-Coombes (1971) agrees that the final fate of the small colony of Solitaires, which had migrated to the proximity of the caves, was probably caused by fire. However, he points out that there were other elements at play before the advent of the fire. He feels that a rapid depletion of the Solitaire population began when the exploitation of the tortoises led to the island being inhabited by sailors and slaves, who would have relished an occasional change of diet from tortoise meat. Cats were also brought in to control an infestation of rats and both of these animals would have caused problems for any ground-dwelling birds. The introduced pigs would have been competing with the Solitaire for the same diet of fruits and roots, and so the birds were

Entrance to Grande Caverne, the François Leguat Giant Tortoise and Cave Reserve.

Latimer

forced to retreat to the less populated, south-western area of the island. The final blow to the Solitaire's delicate existence probably occurred when a fire overtook the region.

The calamity happened at great speed; it was something that would have overtaken the tortoises and killed them on the spot, leaving no evidence of their demise. The Solitaires were able to run away from the hot flames to the cool caves where the dense smoke followed them, pushing them into the darkness where they scrambled into any crevice to escape the suffocation that was to come.

Grande Caverne, the François Leguat Giant Tortoise and Cave Reserve.

So when did this event take place on Rodrigues? Hachisuka (1953) writes that:

> *Mr. Gorry, who lived in the island forty years, never saw a bird as large as that indicated by excavated bones. This would indicate that by 1791, a hundred years after Leguat arrived at the island, no traces of living Solitaires were to be found. Since we only have Leguat's testimony, the bird may have become extinct long before this. The Abbé Pingré, however, who visited Rodriguez in 1761, to observe the famous Transit of Venus, states he was assured that the Solitaire was then alive.*

Fuller (2001) observes that '*although the species seems to have been still common enough during the 1720s, it must have declined rapidly after this … so it seems the species died out during the second half of the eighteenth century.*'

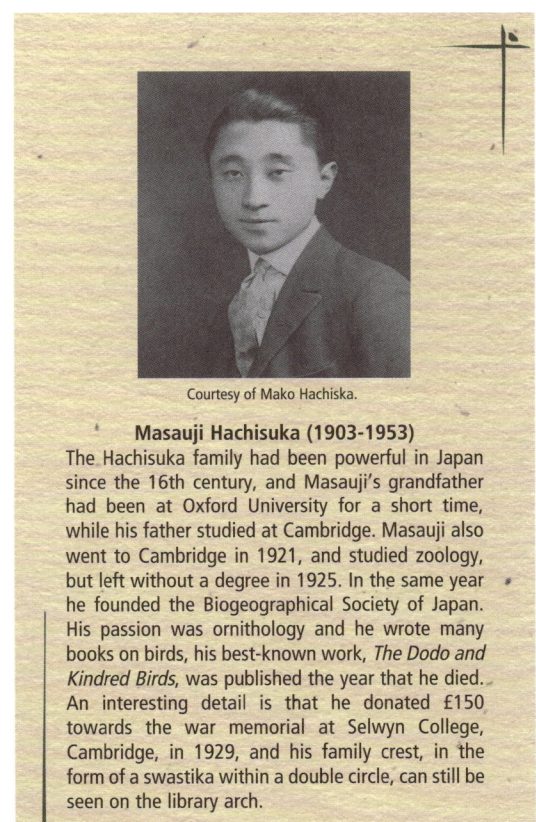

Courtesy of Mako Hachiska.

Masauji Hachisuka (1903-1953)
The Hachisuka family had been powerful in Japan since the 16th century, and Masauji's grandfather had been at Oxford University for a short time, while his father studied at Cambridge. Masauji also went to Cambridge in 1921, and studied zoology, but left without a degree in 1925. In the same year he founded the Biogeographical Society of Japan. His passion was ornithology and he wrote many books on birds, his best-known work, *The Dodo and Kindred Birds*, was published the year that he died. An interesting detail is that he donated £150 towards the war memorial at Selwyn College, Cambridge, in 1929, and his family crest, in the form of a swastika within a double circle, can still be seen on the library arch.

View of Leguat's settlement, from the original in Strickland and Melville (1848). Pencil drawing by J.-F. Sookahet.

Auis Indica

9 Kindred animals

As well as hosting the Solitaire, Rodrigues was home to many other interesting animals and birds, many of them now extinct, but some are still holding on.

Rodrigues has changed a great deal since the time when François Leguat arrived with his companions to gaze at the low mountains, richly covered with beautiful trees, and to marvel at the streams descending to the valleys below, before finally flowing out to sea.

The group found many intriguing animals including large tortoises, which were so numerous *'that sometimes you see two or three thousand of them in a flock: so that you may go above a hundred paces on their backs … without setting foot on the ground'* (Leguat 1708). With this number of tortoises it was natural for Leguat to write about them, which he did in the most complete and accurate way. He had no previous reference material and his observations of the tortoises and the sea turtles were not influenced by the experiences of others.

The largest tortoises surviving to this day are found on the Galapagos Islands (west of Ecuador) and the Aldabra Islands (south-west of the Seychelles). Species of these tortoises were originally found on Madagascar and the Comoros, Seychelles and Mascarene islands. They disappeared from Mauritius and Réunion in the early 1700s, and tortoises were last seen in Rodrigues in 1795. Two species of giant tortoise were found on Rodrigues: the Domed Tortoise and the larger Saddleback Tortoise.

Latimer

Above: Rodrigues Saddleback Tortoise and Leguat's Gelinote. Wall painting at the Forestry Quarters, Rodrigues by Julian Hume.
Left: Some extinct animals and birds of Rodrigues. Composition by J.-F. Sookahet.

Giant tortoises, like those found on Rodrigues, thrived on remote islands where there were no predators. Here they relaxed their defences by increasing the shell openings to enable greater agility and feeding range. This modification would have been a certificate of death in countries where lions and other predators would have taken advantage of such exposure. Another adaptation, which all the Mascarene tortoises seem to have made, was to dismantle the protective features of their shells, making them thinner and lighter. This enabled them to move a little faster over the ground, so that they became the 'sprinting tortoises'!

A strange habit recorded by Leguat, was that the tortoises seemed to have what appeared to be sentinels at the four corners of a herd. These sentinels placed themselves a little distance from the others, facing outwards as if keeping watch – but what they were looking for, and what they would do if they saw danger, no one knows. Unfortunately, these animals were incapable of defending themselves and they had no hope of escape.

In a way, it was fortunate for the Solitaires that the tortoises were so large that they could not help taking centre stage when visitors first stepped onto the island. The fact that they were large and slow moving made it a certainty that they would be easy prey for hungry sailors looking for a quick meal. It seems that these animals made no attempt to hide, and during the act of mating the triumphant roar of the mounted male could be heard over long distances, as this usually rock-like animal showed much passion and animation. The result of this union came in the form of eggs, and Leguat tells us that *'each female lays 1000 to 1200 eggs every year, the biggest of them laying 200 on several occasions; in less than two hours.'*

One can easily imagine why the tortoises were forced into extinction, as they were almost made for this purpose:

> *Of all the creatures that have lived on earth, surely few are more unlikely candidates for survival than tortoises. I use the word 'unlikely' because the natural world has always been a competitive one, murderous on those who fail to meet its standards. It is difficult to conceive that a creature with no offensive capability, with locomotion so compromised by the massive body and the awkward stance of the limbs that any predator could capture it without even giving chase, and that offers such a predator a substantial quantity of excellent meat, could have survived so long.*
>
> (Pritchard, 2005)

The probability of the Solitaire surviving the human onslaught was much higher than that of the tortoise. Although it did not fly, it had a reasonable view of the world around it as it looked down onto the rock-like backs of the tortoises, which spread like tiles across the countryside. The tortoises did not have such a good view of their habitat as their low-slung bodies only gave them a limited viewpoint of the world, dominated by details of a metre or so of ground, and low vegetation.

http://www.furry.org.au/kangaroos/
Courtesy of Marko Laine.

Harriet the tortoise dies aged 175

Harriet the tortoise, one of the oldest animals in the world who some claim was studied by the pioneering 19th century naturalist Charles Darwin, has died in Australia at 175 years of age. The giant Galapagos tortoise passed away from heart failure after a short illness, according to an Australian vet. The 330 lb tortoise hatched on the Galapagos Islands in 1830, but lived at Australia Zoo in southeast Queensland during the final 17 years of her life.

Harriet, who was originally mistaken for a male tortoise and called Harry for more than a century, was named the world's oldest living animal in the Guinness Book of Records. When she was DNA tested, scientists were able to say she was born around 1830 – five years prior to Darwin's visit to the Galapagos archipelago. In 1835, he took several young giant Galapagos tortoises back to London to study on board HMS Beagle, but researchers later were of the belief Harriet belonged to a sub-species of tortoise only found on an island that Darwin had not visited.

'It's thought she may have been taken off there (Galapagos) by Charles Darwin,' John Hangar said. 'She's spent a period of time in Britain and found herself at the Botanic Gardens in Brisbane from about 1850 or 1860 onwards and eventually she found her way up to Australia Zoo.'

by Lee Glendinning and Agencies
The Times, 23 June, 2006

As well as land tortoises, Leguat and his companions noticed that *'The sea-turtles are prodigious big; we have seen some that weighed above 500 pounds.'*

When Leguat looked down from the rocks into the clear pools of the sea he would have seen many fish, as well as a striking animal called a Lamantin. This sea mammal had a head that looked like that of a pig, two paw-like fins and a tail. They could measure up to 3 or 4 m long and weighed around 400 kg. Known now as a Dugong or Sea Cow, Leguat described it accurately:

The body is quite big up to the naval, and the tail is similar to that of a whale, which is horizontal when it lies on its stomach. It is warm-blooded, its skin being dark, very rough and hard, with some hairs, but so sparse that you can hardly see them. Its eyes are small, and the two orifices that it opens and closes can be justifiably called its gills and ears. As it often has its tongue within, which is short, many think that it has none. It has hind-teeth and even tusks like that of a boar, but not fore-teeth. Its gums are strong enough to tear and chew grass, its flesh being excellent, tasting almost the same as veal. It is very wholesome meat.

The female has breasts like that of a woman. Some say she brings forth two offspring at a time, giving them milk together, and carrying them both against her breasts with both paws. But since I have never seen her carry but one, I am inclined to believe that she only bears one young at a time.

These animals were easy to catch as they allowed the settlers to prod and poke them in order to select a good one for eating. The selected Dugong was caught by putting a rope around its tail and simply dragging and carrying it out of the water. Tafforet (1726) wrote that *'Lamantins are abundant at*

Rodrigues especially during the mating season. I have seen schools of 30 to 40 grazing the seaweeds at depths of 2 to 3 feet. Some are 15 to 18 feet long.'

By the time that Pingré visited in 1761, the Dugong had become much rarer, *'Yesterday while sailing in the roadstead we saw a few Lamantins.'* Today, Dugongs are not quite extinct, although they are not seen around the Mascarene Islands anymore. There has been a drastic drop in the world population since the 1950s, and these rare mammals are seen only around islands off India, Singapore, Malaysia, Indonesia, Australia and East Africa.

While wading or swimming in the sea, Leguat mentions that he was often surrounded by a large school of sharks, which did not seem aggressive towards him,

although they were nearly 5 m long and had several rows of sharp teeth.

Another animal observed by Leguat, which has managed to survive until today, is the Rodrigues Fruit Bat. However, it is on the critically endangered list with an estimated 5000 individuals left in the wild. The major threats to its existence are the deforestation of Rodrigues, which leaves them homeless, and cyclones, which blow down the remaining trees and strip away the fruits upon which they feed.

These bats are covered with dark-brown fur, except for a golden colour around the head, neck and shoulders. They have a wingspan of almost a metre, have no tail and weigh around 450 g. Because of their dog-like faces, large eyes and widely spread ears,

Dugong or Leguat's Lamantin (Le Lamentin), from his *Voyage et Avantures* (1708). Courtesy of Owen Griffiths.

Rodrigues Fruit Bat. Courtesy of the Mauritian Wildlife Foundation.

they are sometimes known as Flying Foxes. Their diet of fruit, which they find by sight as well as smell, is high above the range once foraged by the Solitaire, so there was no competition. Indeed, their habit of dropping fruits probably benefited the Solitaires. Unlike the Solitaire they live in colonies; the females roost in groups, whereas the males roost by themselves, marking their territory by rubbing their heads, necks and shoulders on the branches.

Leguat was the first to mention that he saw these bats flying in the daytime, and at this time European travellers had not recorded bats that were active during the day – another instance where Leguat's testimony has since been proven to be correct. Thomas Corby, a government surveyor, also confirmed this habit on a short visit to Rodrigues in 1845, by observing that *'we went to Oyster Bay (from Northern Bay) … in the course of the day we saw some Flying Foxes.'*

In 1963, Gerald Durrell established the Jersey Wildlife Preservation Trust in order to breed certain endangered species in captivity within the country of origin. The Rodrigues Fruit Bat was one of the first animals to be considered for this custodial protection:

At first glance through the binoculars it appeared that each mango tree had produced a strange crop of furry fruit, chocolate brown and golden red. But as the bats yawned and stretched you could see the leather, umbrella-like wings were dark chocolate brown, while the fur on the bodies and heads ranged from bright, glittering yellow, like spun gold, to a deep fox red. They were, without doubt, the most colourful and handsome fruit bats I had ever seen. They had rounded heads with small, neat ears and short, somewhat blunt muzzles that made them look like Pomeranians. The bulk of the colony hung in those three mango trees, and solitary individuals roosted in the smaller trees around.

(Durrell, 1977)

Leguat also described the Rodrigues Giant Day Gecko (previously called the Rodrigues Giant Lizard) with its array of many colours:

> *The palm trees and lataniers are laden with foot-long lizards, the beauty of which we could not stop admiring. There are black, blue, green, red and grey, all of the brightest and liveliest colours.*

These geckos seemed to be quite tame and would eat fruit from the table and even from Leguat's hands. He described their size as being *'as long as an arm'*, and later Vinson and Vinson (1969) wrote that *'the total length of the Rodrigues Giant Lizard must be of the order of 440 mm in fully grown individuals.'*

Another smaller gecko was later called Newton's Day Gecko. It reached a known length of 23 cm, although one historical account places the maximum length at up to 27 cm. Like most other *Phelsuma*, this gecko was capable of rapid colour changes and it fed on insects, pollen and fruit. All that remains of this gecko species today are six preserved specimens, three of which are in the Natural History Museum, London, and three others are in the Paris Museum. The lizard occurred on the mainland of Rodrigues and on the offshore islets. Extinction was probably complete on the mainland by the 1870s and the gecko was last collected on one of the small, offshore islets in 1917. Although fairly thorough searches were conducted in the 1960s and 1970s on all offshore islets they failed to yield any signs of this lizard.

Leguat also reported seeing some small pigeons, and their presence on the island at that time was confirmed by Milne-Edwards (1896), when he examined excavated bones now at the Paris Museum. Leguat probably described the Rodrigues Pigeon:

> *The Pigeons here are somewhat less* [in size] *than ours* [in Europe] *and all of a slate colour, fat and good. They perch and build their nests upon trees; they are easily taken, being so tame, that we have had fifty about our table to pick up the melon seeds which we threw them, and they liked mightily. We took them when we pleased, and tied little rags to their thighs of several colours, that we might know them again if we let them loose. They never missed attending our meals, and we called them our chickens. They never built their nests in the Isle, but in the little Islets that are near it. We supposed it was to avoid the persecution of the rats, of which there are vast numbers in this island.*

Newton's Day Gecko. Acrylic on paper by Julian Hume.

Rodrigues also had its own specific *Aphanapteryx,* the twin of the extinct rail found in Mauritius. This was Leguat's Gelinote (an old French word for hen, which in English could be translated as wood-hen):

Our Wood-hens are fat all the year round and of a most delicate taste. Their colour is always of a bright grey, and there is very little difference in the plumage between the two sexes. They hide their nests so well that we could not find them out, and consequently did not taste their eggs. They have a red list upon their eyes, their beaks are straight and pointed, near two inches long, and red also. They cannot fly as their fat makes them too heavy for it. If you offer them anything that's red, they are so angry they will fly at you and catch it out of your hand, and in the heat of the combat we had an opportunity to take them with ease.

(Leguat, 1708)

Rail bones have been found and examined by Milne-Edwards and connected with Leguat's description. Leguat was probably describing a female where the straight bill is a sex characteristic, and fits in with the drawing of the Red Rail by Herbert (1638), when he saw the same bird in Mauritius.

Tafforet also described the bird, but stated that the beak was curved like that of a curlew. This agrees with osteological evidence, which shows that the beaks varied in their curvature.

One of the birds mentioned by Leguat and his companions was the Paille-en-queue

(White-tailed Tropicbird), as these birds used to swoop down and try to steal their caps. They can still be seen today with their long tails, hovering and swooping over trees and rocks.

Leguat and Tafforet both make passing reference to an owl endemic to Rodrigues but unfortunately without much detail. This was probably the Rodrigues Owl, and later on bones confirming the presence of an owl were found on the island. Pingré does not make any reference to owls during his visit and it is assumed that the owl had become extinct before 1761.

Leguat's Gelinote imagined by F. W. Frohawk, from Rothschild's *Extinct Birds* (1907).

Skulls of a parrot have been found on Rodrigues, which could be from the long-tailed, large-headed, green parrot described by Tafforet (1726). It seems that this parrot, the Large Green Parrot, populated the islets off the south of the main island, where it fed

Rodrigues Night Heron. Wall painting at the Forestry Quarters, Rodrigues by Julian Hume.

on small black seeds from a tree with leaves smelling of lemons, and had to fly across to the mainland to find fresh water.

A bird that was warmly recorded by Leguat was later named Newton's Parakeet. Alfred Newton was the first scientist to describe it from bones deposited at the University Museum of Zoology at Cambridge, where he was professor of zoology. Leguat and his friends taught a few of these birds to speak, and they took one of them on their journey from Rodrigues to Mauritius. Despite the fact that some of the birds were treated as pets this did not stop the band of castaways from eating the rather delicious flesh.

In their writings, both Leguat and Tafforet agreed that they saw a number of bold and confiding birds the size of egrets, which, although they could run well, were very poor at flying. Gunther and Newton (1879) examined material buried on the island and confirmed that the remains found from these birds belonged to the Rodrigues Night Heron.

Another bird seen and described by Tafforet (1726) was the Rodrigues Starling. Bones that could have come from the same bird were discovered by the Revd Slater in 1874:

A little bird is found which is not common … one sees it on the Islet aux Mât, which is in the south of the main island, and I believe it keeps to that islet on account of the birds of prey which are on the mainland, and also to feed with more facility on the eggs of the fishing birds which feed there, for they feed on nothing else but eggs, or some turtles dead of hunger, which they … tear out of their shells. These birds are a little larger than a blackbird, and have white plumage, part of the wings and tail black, the beak yellow, as well as the feet, and make a wonderful warbling. I say a warbling since they have many and altogether different notes. We brought up some with cooked meat, cut up very small, which they eat in preference to seeds.

We have to marvel at the proven accuracy of Leguat's descriptions of the flora and fauna he found around him, and because of this it would be wrong to omit his sighting of a bird he called the Giant, which he mentioned when he was in Mauritius, but records the fact that he first saw one in Rodrigues:

> One sees large flocks of bitterns and great numbers of a kind of bird we called Géants because their heads rise to about six feet high. They are very tall when they stand up, and their neck is very long. The body is about the size of a Goose. They are completely white except for a red spot under the wing. They have the beak of a Goose, but a bit more pointed, and their toes are separated and very long. They pasture in marshy places, and the dogs often take them by surprise as they take some time to fly away. Once we saw one on Rodrigues, and it was so fat that we caught it with our hands. It was the only one we saw, which made me believe that it had been driven by some wind that it could not resist. Its meat was quite good.

It has been suggested that Leguat was actually describing a flamingo. However, since he had studied and sketched the flamingo whilst in the Netherlands, he

White-tailed Tropicbird. Photograph by Dr Nik Cole.

would have known that these birds usually have white bodies with red wings tipped with black and he would have recognised it as such. Also, no one in the group that saw the stray bird in Rodrigues suggested that it was a flamingo. The dogs that chased the birds were obviously those kept by the Dutch either at Black River or Flacq in Mauritius, where the Huguenots stayed for some time. These areas were flat and interspersed with shallow marshes, which were filled in during the anti-malarial campaigns.

Flamingo bones were found at Mare aux Songes by George Clark in 1865, but, so far, no bones of the Géant have been found,

even in the more recent Dodo Research Programme. Because of this, many naturalists still think that Leguat's Géant was a flamingo. Even Strickland and Melville (1848) noted *'the fact is that these Géants are evidently Flamingos.'* Both Schlegel (1866) and Oustalet (1896) proposed that it was a Giant White Water-hen, and Rothschild (1907) adds his support by suggesting that *'Leguat's figure and description cannot be meant for a Flamingo and they prove the former existence of a gigantic Ralline bird* [family includes rails, crakes and coots] *in Mauritius.'*

The distinguished Leiden professor, H. Schlegel (1866), when writing on the subject

The Giant (Giant White Water-hen), here labelled *Avis Indica*, found in Hachisuka (1953). Picture by the Flemish engraver, Adriaen Collaert (born 1520), and later described by Leguat who used this engraving for his own drawing of the bird.

of the Géant, commented that *'Hamel and Strickland are, so far as I know, the only persons who have offered their opinions on the subject.*

Stamp showing Rodrigues (or Newton's) Parakeet. Courtesy of the Philatelic Museum, Mauritius.

They had not the least doubt as to the existence of this large animal; nor can such be possible since the accounts of Leguat are too precise.' In fact, Schlegel thought very highly of Leguat, describing him as *'a little known man, who has deserved the thanks of science.'*

A list of some animals and birds (now extinct) that once lived on Rodrigues can be found in *Appendix F*.

Extinct birds and animals of Rodrigues by Julian Hume, at the François Leguat Giant Tortoise and Cave Reserve Museum, Rodrigues. Courtesy of Owen Griffiths. Photograph by Elizabeth Weaver.

10 The illusive Dodos

For many years ornithologists have argued that two separate species of Dodo lived on the island of Réunion, the largest island of the Mascarene group. They named these illusive Dodos the Réunion Solitaire and the Réunion White Dodo. Recently, it has been proposed that the 'Solitaire' was actually an extinct ibis (Mourer-Chauviré et al., 1995), and that the 'White Dodo' was brought into existence by artists copying a Roelandt Savery painting.

(Hume and Cheke, 2004; Valledor de Lozoya, 2003)

EXTINCT BIRDS

PLATE 25

DIDUS SOLITARIUS
(One-Third Natural Size—*from a Dutch picture taken from living bird in Amsterdam, beak and wing restored*)

It is natural to believe that as there was once a Dodo living on Mauritius, and a Dodo-like cousin living on Rodrigues, then there must have been a similar species living on Réunion. This island lies only 160 km to the south-west of Mauritius, giving over much of its terrain to rugged mountains covered by forests, ringed by fertile coastal lowlands. Some 100,000 years

Above: The White Solitaire of Réunion, *Didus Solitarius*, cromolithograph by F. W. Frohawk (1907). Courtesy of IPC Ltd.
Left: An improbable Dodo by Hoefnagel (c. 1600) and an illusive White Dodo by Holsteyn (c. 1630) in a composition by J.-F. Sookahet.

ago, one of its volcanoes, Piton des Neiges, had a catastrophic eruption, killing off most of the flora and fauna of the island. Now extinct, it is the highest peak of Réunion, rising to an elevation of 3,070 m.

Certain sub-fossil bones of birds such as flamingos, shell-ducks, night herons, kestrels, coots and owls have been found. However, no Dodo or Solitaire bones have ever been discovered. There are eyewitness accounts that these two birds were on the island, but we have to acknowledge that some of the descriptions may have applied to other birds, until material evidence is found.

The evidence for the existence of the Réunion Solitaire is very limited and relies on a few eyewitness accounts, paintings and drawings, many of which were mere copies of previous works, but for the sake of completeness these descriptions are given here.

In 1699, Abbé Carré, who was the secretary of the French East Indies Company, described a bird he called the *'Oiseaux Solitaire for to be sure, it loves solitude and only frequents the most secluded places; one never sees two or more together; it is always alone. It is not unlike a turkey, if it did not have longer legs. The beauty of its plumage is a delight to see. It is of changeable colour which verges upon yellow.'*

The Abbé also mentioned that *'we wished to keep two of these birds to send to France and present them to His Majesty, but as soon as they were on board ship, they died of melancholy, having refused to eat or drink.'*

It was reported that Governor Labourdonnais sent a Solitaire to France as a present for the director of the French East Indies Company, sometime between 1735 and 1746. It is not certain where this bird was from, but it must have died on the journey, as there are no records of it arriving.

Sieur Dubois (1674), a French colonist sent to Réunion from Madagascar, stated that *'These birds are so called* [Solitaires] *because they always go alone. They are as big as a large goose and have white plumage with the tips of the wings and tail black. The tail feathers are like those of the ostrich, they have a long neck and … the legs and feet are like those of the turkey. This bird is caught by running after it, since it can scarcely be said to fly at all.'*

According to Mourer-Chauviré *et al.* (1995) it is almost certain that all the descriptions of the so-called Réunion Solitaire actually referred to what is now known as the extinct Réunion Flightless Ibis. The name 'Flightless' Ibis is misleading, since eyewitness reports indicate that it could actually fly some distance after a running take-off. Bones have been found to prove the existence of an ibis in Réunion, and until bones belonging to the Solitaire are found, we must believe that the source of the name Solitaire actually came from the Réunion Flightless Ibis, with its closest relative being the Sacred Ibis from Africa.

One of the rare images of the other Dodo supposed to have lived on Réunion, the Réunion White Dodo, is portrayed in an oil

Reconstruction of the Réunion Ibis by Julian Hume. Courtesy of IPC Ltd.

painting by Roelandt Savery. It is entitled *Orpheus Charming the Animals* and is currently on display at the Gemäldegalerie Alte Meister in Berlin. Savery painted many Dodos, but only one white one, and this is only about 1 cm high. According to Arturo Valledor de Lozoya there are several hypotheses as to why this particular painting showed the Dodo as white in colour, but the most probable is that it was merely artistic licence:

> ... white Dodos could have had a real existence as immature specimens of the brown [sic] *Dodo of Mauritius, but not as a separate species present in Réunion until its extinction.* (Valledor de Lozoya, 2003)

Much of the research on the Réunion White Dodo is based on the work and interest of Lord Walter Rothschild who published his book *Extinct Birds* in 1907. Lord Rothschild came from a family of wealthy bankers but his consuming interest was in ornithology, and extinct birds in particular. He placed his many paintings and stuffed reconstructed birds in the museum he founded at Tring in southern England.

Rothschild's main interest was taken up by Hachisuka (1953), who maintained that there must have been two Dodo-like creatures in Réunion: a White Dodo and a White Solitaire. He based his theory on various reports and pictures.

Following a voyage undertaken in 1613, John Tatton, who was an officer serving under Captain Samuel Castleton on the ship *Pearl*, gave his account of the White Dodo (1625). He described it as '*a great fowl the bigness of a Turkie, very fat, and so short-winged that they cannot flie, being white, and in a manner tame; and so are all other fowles as having not been troubled or feared with shot.*'

Fuller (2002) made the following observation:

> If a Dodo-like creature did, in fact, once inhabit Réunion it would certainly not have looked exactly like a regular Dodo, with the only major difference lying in the colouration of the plumage. Any creature evolving on Réunion would have been evolving—once it had lost the power of flight—in complete isolation. It would, therefore, have followed a rather different line to that pursued by the Dodo of Mauritius and would certainly have acquired a number of independent features.

The differences between the Mauritian Dodo and the Rodrigues Solitaire clearly demonstrate that parallel evolution on different islands results in important morphological differences. If a Dodo species did live on Réunion, it almost certainly would have looked quite different from its Rodrigues or Mauritian cousins.

White Dodo with other birds. Watercolour by Pieter Withoos (c. 1684). Courtesy of IPC Ltd.

Fabrice Desvoux

11 The Solitaire today

It has been an opinion, that he who receives an Estate from his ancestors is under some kind of obligation to transmit the same to their posterity.   (Franklin, 1789)

After the demise of the Solitaire it was largely forgotten, but scientists and ornithologists have gradually pieced together enough material evidence to enable us to have a fair idea of its appearance and lifestyle. Great efforts are now being made to protect the rich biodiversity of Rodrigues, and it is hoped that residents and visitors alike can benefit from the inheritance that the Solitaire has left behind on the island.

This chapter is in two main sections. Firstly, it details local sites related to the history of the Solitaire and the island on which it lived, and highlights conservation initiatives/ organisations working to protect the native flora and fauna of Rodrigues. Secondly, it enables readers to share in some of the modern Solitaire experience by showing, pictorially, the presence of the Solitaire in everyday life in Rodrigues.

Local sites, initiatives and organisations

The Grande Montagne Visitor and Information Centre

This is a good starting point for those who want to find out more about the Solitaire. Here visitors can see:

- An almost complete skeleton of the Solitaire from the bones collected by Jean Richard Payendee (Mauritian Wildlife Foundation) from Caverne Poule Rouge in the year 2000
- Various Solitaire pictures
- A carapace of a giant land tortoise.

Above: Grande Montagne Visitor and Information Centre.
Below: Inside the Centre.

Left: Pencil drawing by Fabrice Desvaux de Marigny (aged 11) – Alexandra House School, Mauritius.

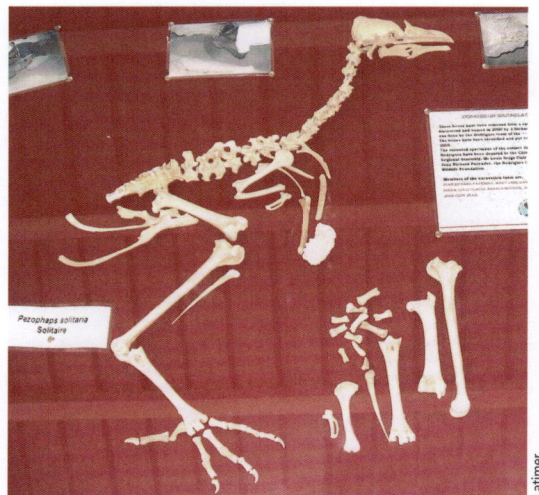

Solitaire skeleton in Grande Montagne Visitor and Information Centre.

Grande Montagne and Anse Quitor nature reserves

These nature reserves are actively involved in conservation work, funded by the World Wildlife Fund (WWF), which began in Rodrigues in 1985. Several areas in the

Young forest in Grande Montagne Nature Reserve.

upland Grande Montagne and lowland Anse Quitor reserves were weeded and planted with native trees. This work was undertaken in collaboration with the Ministry for Rodrigues Forestry Service. The focus was on rare plant conservation and the planting of native trees.

In 1996, funding was secured from the World Bank/Global Environment Facility (GEF). The focus during this programme has been to work in areas of forest with a remnant of native vegetation and to use these as core areas around which to eventually build a fully functional native forest. The work is based at Anse Quitor and Grande Montagne and involves the full collaboration of the Forestry Service.

The first phase of this programme consisted of weeding the exotic vegetation in selected plots in the reserves followed by the planting of native pioneer species. Once good cover has been secured, native climax[6] species will be planted. Experimental plantings are determining the best species to use as pioneers, as well as the optimum shading conditions for the success of transplanted seedlings.

In the future it is planned to experiment with different weeding techniques and to use rat control to enhance the natural regeneration work. This will enable the restoration of larger areas.

---

[6] Plants that have remained stable under the prevailing environmental conditions.

The project is also propagating and replanting very rare plants in order to secure the future of many species and the genetic variability within them.

There are other nature reserves on Île Cocos and Île aux Sables, off the north-west coast of Rodrigues.

Whilst walking around the various nature reserves and sites in Rodrigues visitors may be interested in the animals and birds that still inhabit the island. A list of some of these may be found in *Appendix G*.

The François Leguat Giant Tortoise and Cave Reserve

This project was initiated by Owen Griffiths, an Australian zoologist, who is reintroducing the giant land tortoises to Rodrigues. The project has three main objectives:

- To establish a *wild* population of giant land tortoises on Rodrigues
- To recreate the native and endemic forest
- To protect the rare biotype of the limestone region, including the caves, which are rich in sub-fossils of the extinct original fauna.

The François Leguat Giant Tortoise and Cave Reserve Museum. Photographs by Elizabeth Weaver.

The site covers an area of 19 hectares and is situated on the limestone plain in the south-west of the island. It is hoped that, within the next few years, the island that François Leguat described will become a reality on this site. In the museum, visitors will have the opportunity to learn more about the extinct fauna of Rodrigues. The site includes Grande Caverne, which is where eminent naturalists dug for Solitaire bones over 130 years ago.

According to historical records, particularly by Leguat and Tafforet, Rodrigues had the densest giant land tortoise population (300,000) in the world. Unfortunately, the Rodrigues tortoise species were completely eradicated and the last record of a living tortoise in Rodrigues was mentioned by Philibert Marragon in 1795. Because the two tortoise species from Rodrigues are now extinct, the island will be populated with similar species, originally collected from the Aldabra Islands and southern Madagascar, and bred in Mauritius. Visitors to the reserve will have the unique experience of walking amongst and observing these giants in their natural environment.

Entrance to a small cave at the François Leguat Giant Tortoise and Cave Reserve.

Solitaire remains

There are a few places on the island where Solitaire remains can be viewed:

- Grande Montagne Visitor and Information Centre
- Mauritian Wildlife Foundation, Forestry Quarters
- The François Leguat Giant Tortoise and Cave Reserve Museum.

Caves

Rodrigues is fortunate in having many interesting caves, but the most accessible are Grande Caverne on the site of the François Leguat Giant Tortoise and Cave Reserve and Caverne Patate.

Examining a trench in Grande Caverne, originally excavated in the 1860s.

The Mauritian Wildlife Foundation

The Mauritian Wildlife Foundation is dedicated to saving the country's endangered wildlife from extinction. With this self-imposed challenge it is one of the leading forces in Rodrigues for the preservation of plants and animals. Visitors can drive through an avenue of trees to find out more at the Forestry Quarters, Solitude.

Eco-tourism

In recent years, a new class of tourist has emerged. This tourist expects to do more than just swim in the sea and sit on the beach. The eco-tourist likes to explore the habitat and environment of a destination, and wants to get to know more about the natural history of the places they visit. Rodrigues is beginning to promote itself as an eco-tourism resort, which has the following attractions:

Rodrigues … offers tremendous possibilities for eco-tourism. Its hilly terrain and sandy

The Résidence, Port Mathurin.

beaches, unspoilt by pollution, make the island of 110 sq. km a hiking country. Most of the walks provide scenery of breathtaking beauty. Its endemic vegetation of fan palm, feather palm and screw pine and other species ranks this island as unique in the Mascarenes. Further, its limestone caverns with stalactites and stalagmites are well worth a visit. Off the coast of Rodrigues are several islets of great beauty. Two of these are sandy cays, Île aux Sables and Île Cocos, and are sanctuaries for birds.

(Éditions le Printemps Ltée., 2001)

The Mauritius Tourism Promotion Authority is active in promoting the island of Rodrigues as an eco-tourism resort. In fact, it has become so popular with visitors that offices have been opened at the Résidence in the centre of Port Mathurin.

Protecting birds in Rodrigues

The Government of Mauritius appreciates the fact that it is essential to preserve the natural habitat, and to protect the bird life that history has proved can so easily disappear. Mauritius can still 'boast' that it has three of the world's most threatened birds: the Mauritius Kestrel, the Echo Parakeet and the Pink Pigeon. A number of organisations, including the Mauritian Wildlife Foundation, are now working hard to preserve the rare birds on Mauritius and Rodrigues.

Protecting the forests in Rodrigues

During the last three centuries, the native forests in Rodrigues have been indiscriminately cleared for agriculture and human settlement. This has resulted in the loss of the ancient forests, and thus the loss of the habitat necessary for hosting indigenous and endemic flora and fauna. Some native species have been irrevocably lost and quite a few are critically endangered. It has been necessary in some cases to fence off small parcels of forest to keep out the goats and cows, and labourers spend weeks uprooting and burning plants that have invaded the reserved land. In certain areas the endemic plants are planted alongside exotic plants where they are expected to fight for their survival – a perfect example of the survival of the fittest. See *Appendix H* for some of the endemic and native coastal tree species being planted around the island.

The Solitaire in everyday life

As we turn our attention to the Solitaire in our modern age, we have to acknowledge that the term has several other meanings. It is not a unique name for the extinct bird which is the subject of this book, there are several other birds called Solitaire, including a dull grey North American thrush noted for its beautiful song.

The word is also used in other contexts with no connection at all with birds:

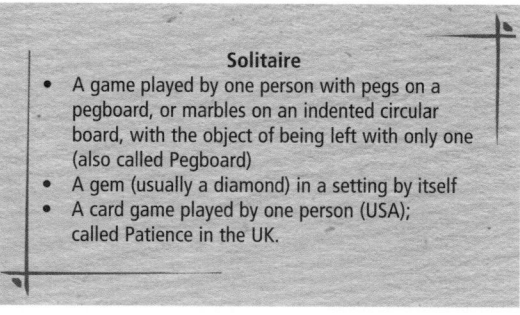

Solitaire
- A game played by one person with pegs on a pegboard, or marbles on an indented circular board, with the object of being left with only one (also called Pegboard)
- A gem (usually a diamond) in a setting by itself
- A card game played by one person (USA); called Patience in the UK.

The Solitaire in everyday life. Photographs by Bob Latimer.

Top left: Inside Artisanat La Colombe.
Bottom right: Model of the Solitaire in Cotton Bay Hotel.

Images and representations of the Solitaire can be seen increasingly on Rodrigues, although nowhere near to the extent that the Dodo can be seen worldwide. With increasing numbers of tourists visiting the island, there is a growing need for good quality Solitaire memorabilia. This can be found in a variety of craft shops, which offer a limited range of Solitaire items.

Artisanat La Colombe. Proprietor – Frankilaine Cesar.

There is a rather unusual craft centre called CareCo where all the craftwork is made by visually and hearing impaired islanders, who initially attended the school attached to the centre. Visitors are always welcome to look around the school and visit the workshops, and then purchase Solitaire artefacts in the knowledge that any profit will fund the education of the children.

Mauritius has a long and interesting philatelic history. The early stamps featured the head of Queen Victoria of England, but since that time special issue stamps have included the Solitaire:

The coat of arms of the Municipality of Rodrigues proudly shows the Solitaire:

As more evidence is gathered on the Solitaire, interest in this bird is bound to grow. Ideally, the publication of this book, and the knowledge that it imparts, will add to this step forward.

1 *Site of François Leguat's hut.*
2 *Pingré's observation point of the transit of Venus.*
3 *Place François Leguat and memorial*
4 *The Résidence.*
5 *Solitaire sculpture at Millennium Square.*

Leguat's settlement is shown in green

Map of Port Mathurin showing the area of Leguat's original settlement and other sites of interest. Composition by J.-F. Sookahet.

These birds of the Mascarene Islands have been saved from the brink of extinction through the hard work of the Mauritian Wildlife Foundation and private individuals who have made a determined effort to do so. Let us add our own personal pledge to support such dedicated conservation work during our short stay on this planet.

Epilogue

In the last 300 years, species of all kinds have been going extinct between five and fifty times faster than before. And the pace is accelerating. (Diamond *et al.*, 1987)

It is hoped that this book will help to elevate the Solitaire to its rightful place, standing proudly and elegantly beside that equally wonderful bird, the Mauritian Dodo. The remoteness of Rodrigues has helped add to the loneliness of the Solitaire and its extinction passed without a flicker of interest from the outside world. Most Solitaires had disappeared by the 1760s, and all that was left were a few bones preserved within the protective environment of the caves and fissures which are still giving up their secrets to persistent scientists.

Although the Solitaire has now departed, it is the only extinct bird to have a place reserved for it in the heavens. After Abbé Pingré visited the island and failed to observe the bird, a French astronomer wrote a paper called the *Constellation du Solitaire* (Le Monnier, 1776) in which he listed twenty-two constituent stars in memory of the voyage to the island of Rodrigues. Unfortunately, the bird in his constellation resembles a Blue Rock Thrush, which was not the species of Solitaire that Pingré had tried to find on Rodrigues.

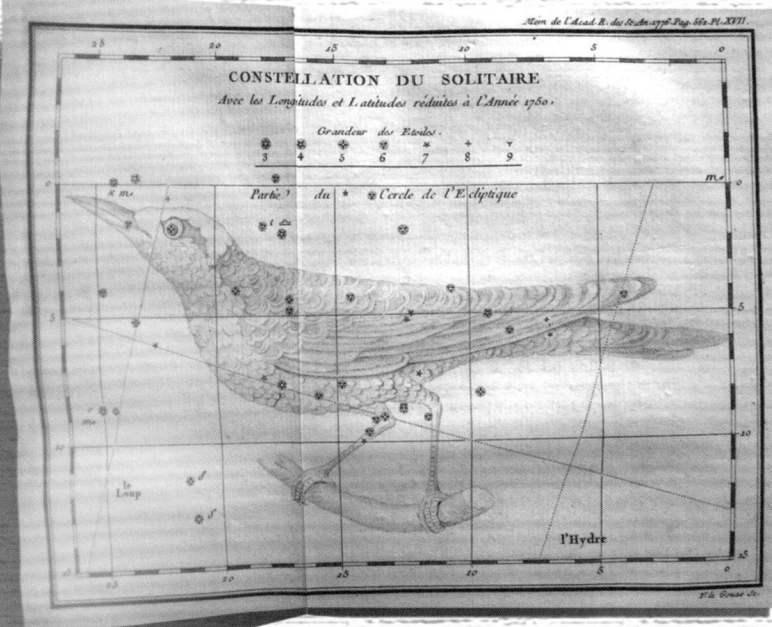

Above: Constellation du Solitaire by Le Monnier, in *Mémoires de l'Académie Royale des Sciences* (1776). Courtesy of Felice Stoppa.
Left: Echo Parakeet, Pink Pigeon and Mauritius Kestrel. Watercolour by Kathleen Latimer.

The lessons we have learned from the demise of the Solitaire, and similar birds, have become a catalyst for a worldwide preservation and conservation order, and there is now a resolution to honour our environmental responsibilities towards our planet.

Present-day bird extinction rates

Every year another species of bird vanishes forever, new research suggests, an extinction rate four times higher than traditional estimates.

Furthermore, the analysis predicts that by the end of the century the rate will grow to ten extinctions per year, meaning the loss of 12% of all 10,000 known bird species.

The work was carried out by Peter Raven, at The Missouri Botanical Garden in St Louis, US, and colleagues …

The researchers argue that the traditional estimate of extinction rate, of one species every four years, does not take into account several crucial factors. These factors include the continual identification of extinct species from skeletal remains – boosting the number of extinct species – and numerous missing species not yet declared extinct.

Additionally, the team notes that extinction-rate estimates are based on data gathered from 1500 to the present day, and because most bird species were first identified between 1850 and 1900, the extinction estimates for the earlier period are underestimated, skewing the data.

Before human activities began to impact on bird survival, the extinction rate would have been about one every 100 years: the natural evolution of at least one bird species per century, Raven says.

But the impact of humans has caused a massive hike in the extinction rate in the islands of the Caribbean, Atlantic, Pacific and Indian oceans. This is mainly due to the destruction of the birds' natural habitats for farmland, but also through hunting and the introduction of alien threats, such as cats and avian malaria.

Extract from an article in *New Scientist* by Gaia Vince (2006).

Albert Pitot (1914) reminds us that respect for the habitat is paramount to the well-being of the animal that dwells there:

> *Each type of animal possesses its own structure, even in the most exceptional and apparently extravagant forms, which corresponds to the conditions and necessities of the environment in which it was destined to live.*

Sadly, the Solitaire is no more, and will never grace the rocks and glades of Rodrigues again. Spooner (1992) puts the extinction of a non-flying bird into perspective by asking us to '*consider the train of events that led to his passing: an avian, possessing the power of flight, finds himself without predators on a luxury island in the Indian Ocean. He no longer needs to fly so he doesn't bother.*' Homo sapiens usually contrives to pass the blame onto the victim of extinction, but each one of us should make our own pledge that the passing of the Solitaire will make us determined that similar conservation tragedies will never happen again.

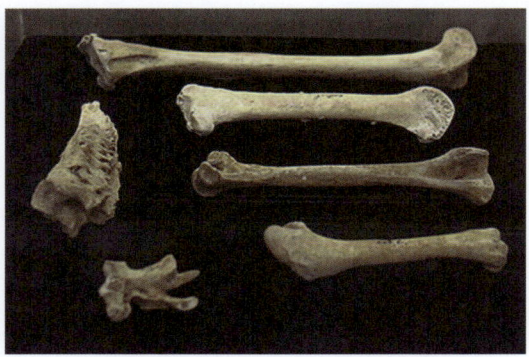

Solitaire bones in the National Museum of Natural History 'Naturalis' in Leiden, Holland. Photograph by Peter Maas (2002).

A Solitaire surveying its territory. Acrylic on paper by Julian Hume.

Pencil drawing of a banyan tree by George André Camille.

Appendix A
Scientific names of animals, birds and plants

Animals and birds

Blue Rock Thrush (constellation) (*Turdus solitarius*)

Blue Rock Thrush (*Monticola solitarius*)

Crowned Pigeon (*Goura victoria*)

Dodo (*Raphus cucullatus*)

Dugong; Sea Cow; Lamantin (*Dugong dugon*)

Echo Parakeet (*Psittacula eques*)

Giant Bird; Le Géant (*Leguatia gigantea*)?

Large Green Parrot; Rodrigues Parrot (*Necropsittacus rodericanus*)

Leguat's Gelinote (*Aphanapteryx (Erythromachus) leguati*)

Mauritius Kestrel (*Falco punctatus*)

Mourning Gecko (*Lepidodactylus lugubris*)

Newton's Day Gecko (*Phelsuma edwardnewtonii*)

Newton's Parakeet (*Psittacula exsul*)

Nicobar Pigeon (*Caloenas nicobarica*)

Pink Pigeon (*Columba mayeri*)

Red Rail (*Aphanapteryx bonasia*)

Réunion Flightless Ibis (*Threskiornis solitarius*)

Réunion Solitaire (*Ornithaptera solitarius*)?

Réunion White Dodo (*Victoriornis imperialis*)?

Rodrigues Domed Tortoise (*Geochelone (Cylindraspis) peltastes*)

Rodrigues Fruit Bat (*Pteropus rodricensis*)

Rodrigues Giant Day Gecko (*Phelsuma gigas*)

Rodrigues Night Heron (*Nycticorax megacephalus*)

Rodrigues Owl (*Mascarenotus murivorus*)

Rodrigues Pigeon (*Columba rodericana*)

Rodrigues Saddleback Tortoise (*Geochelone (Cylindraspis) vosmaeri*)

Rodrigues Starling (*Necropsar rodericanus*)

Rodrigues Warbler (*Acrocephalus rodericanus*)

Sacred Ibis (*Threskiornis aethiopica*)

Solitaire (*Pezophaps solitaria*)

Tooth-billed Pigeon (*Didunculus strigirostris*)

White-tailed Tropicbird; Paille-en-queue (*Phaethon lepturus*)

Plants

Banyan tree (*Ficus benghalensis*)

Bois Pipe (*Dombeya rodriguesiana*)

Bois Puant (*Foetidia mauritiana*)

Café Marron (*Ramosmania rodriguesii*)

Casuarina; Filao (*Casuarina equisetifolia*)

Ebony (*Diospyros diversifolia*)

Latanier Palm; Latanier Jaune; Yellow Latanier (*Latania verschaffeltii*)

Poison Palm (*Hyophorbe verschaffeltii*)

Screw Pine; Pavillon; Pandanus (*Pandanus heterocarpus*)

Stinkwood tree; Bois Cabris (*Clerodendrum laciniatum*)

Tambalacoque tree (*Sideroxylon grandiflorum*)

Appendix B
Names given to Rodrigues throughout history
(According to Millard, 2004)

Date	Name	Date	Name
1502	Dina Moraze	1691	Petite île de Diego Rois
1507	Domingo Friz	1700	Bygarroys
1528	Rodriguez	1725	Île de Diego Rodriguez
1540	Île Roiz	1725	Marianne
1611	San Roderigo	1726	Île Rodrigue
1634	Dygarroys	1809	Île de Rodrigue
1638	Île de Diego Rois (Diogo Roiz)	1810	Rodrigues

Appendix C
Other birds called 'Solitaire'
Medium-sized, mostly insectivorous birds in the thrush family (Turdidae)

Common name	Scientific name
Andean Solitaire	*Myadestes ralloides*
Black Solitaire	*Entomodestes coracinus*
Black-faced Solitaire	*Myadestes melanops*
Blue Rock Thrush	*Monticola solitarius*
Blue Rock Thrush (in constellation)	*Turdus solitarius*
Brown-backed Solitaire	*Myadestes occidentalis*
Cuban Solitaire	*Myadestes elisabeth*
Rufous-brown Solitaire	*Cichlopsis leucogenys*
Rufous-throated Solitaire	*Myadestes genibarbis*
Slate-coloured Solitaire	*Myadestes unicolor*
Townsend's Solitaire	*Myadestes townsendi*
Varied Solitaire	*Myadestes coloratus*
White-eared Solitaire	*Entomodestes leucotis*

Names for the Solitaire in different languages

Language	Name
Dutch	Rodriguez Solitaire
English	Rodriguez Solitaire, Rodrigues Solitaire
French	Dronte de Rodrigues, Solitaire de Rodrigues
German	Rodrigues-Solitär, Rodriguez-Einsiedler
Italian	Dodo di Rodrigues
Spanish	Solitario de Rodríguez

Appendix D
Physical and skeletal differences between the Solitaire and the Dodo

Physical differences

Area of difference	Rodrigues Solitaire *Pezophaps solitaria* (Gmelin, 1789)	Mauritian Dodo *Raphus cucullatus* (Linnaeus, 1758)
Weight	Ave. 24.4 kg (male) [Livezy, 1993]	Ave. 14.3 kg [Grihault, 2005]
Height	85–90 cm	70–75 cm
Colour – males	Grey and brown	Grey
Colour – females	Blonde to brown	Grey
Head	Headband above beak (females)	Bare area around beak
Beak	Slightly curved (male hooked)	Much larger, curved upper beak
Eyes	Black and lively	Yellow
Wings	Used for defence and calling	No special use
Defence	'Musket ball' wing-tip	Powerful beak and feet
Stance	Elegant	Awkward

Skeletal differences

Area of difference	Rodrigues Solitaire *Pezophaps solitaria* (Gmelin, 1789)	Mauritian Dodo *Raphus cucullatus* (Linnaeus, 1758)
Neck	Much longer	Much shorter than Solitaire
Vertebrae	Generally larger	Generally smaller than Solitaire
Ribs	Thicker rounded nodules at ends	Thinner nodules at ends
Pelvis	Very rounded like most pigeons	Unlike other pigeons
Posterior (tail area)	Unlike any known bird	A tuft of frail feathers
Sternum	Large keel, for attachment of pectoral muscles (used for fighting)	Roughly the same, but has five articular surfaces for sternal ribs
Coracoid	Much stouter	Smaller than Solitaire
Wing	Much stouter and has a bony knob	No bony knob
Leg bones	More strongly developed ridges and muscular impressions	Strong legs
Skull	Cranium is narrower and longer; flat on top	Peculiar frontal protuberance
Upper mandible	Axes of nasal process and the maxillary diverge; straight	Axes of nasal process and the maxillary converge; curved
Maxilla	Relatively small	Much larger than Solitaire

Appendix E
Museums with Solitaire material

Museum	Material
Andersonian Museum, Glasgow	Right and left femur, with a fragment of a third, a left tibia, and a right and left tarsometatarsus collected by Telfair
François Leguat Giant Tortoise and Cave Reserve Museum, Rodrigues	The museum and caves provide plenty of material
Grande Montagne Visitor and Information Centre, Rodrigues	An almost complete skeleton from Caverne Poule Rouge, Rodrigues
Hunterian Museum at the Royal College of Surgeons, London	Male and female skeletons – collected in 1874 by the Revd H. H. Slater
Museum of Comparative Zoology, Cambridge, Massachusetts	One fairly complete skeleton
National Museum for Natural History, Stuttgart	Tibiotarsus and tarsometatarsus
National Museum of Natural History, Paris	Femur, tibiotarsus, tarsometatarsus, humerus, vertebrae
National Museum of Natural History, Washington, DC	Some thirty individual bones
Natural History Museum, London	Two complete skeletons and a nearly complete set of bones of two other individuals
Natural History Museum, Port Louis, Mauritius	A complete skeleton
University Museum of Zoology, Cambridge, UK	Complete skeletons of male and female

Appendix F
Extinct animals and birds from Rodrigues

Extinct animals

Common name	Scientific name	Notes
Newton's Day Gecko	*Phelsuma edwardnewtonii*	Searches for insects on trees, also eats pollen and nectar
Rodrigues Domed Tortoise	*Geochelone (Cylindraspis) peltastes*	Carapace length 46 cm
Rodrigues Giant Day Gecko	*Phelsuma gigas*	About 44 cm long
Rodrigues Saddleback Tortoise	*Geochelone (Cylindraspis) vosmaeri*	Carapace length 85-120 cm

Extinct birds

Common name	Scientific name	Notes
Giant Bird (Le Géant)	*Leguatia gigantea*	Not known if it existed
Large Green Parrot	*Necropsittacus rodericanus*	Appearance not known but probably a uniform green colour. Length 50 cm(?)
Leguat's Gelinote	*Aphanapteryx (Erythromachus) leguati*	Also known as the Red Hen
Newton's Parakeet	*Psittacula exsul*	Length up to 40 cm
Rodrigues Night Heron	*Nycticorax megacephalus*	Closely related to Mauritius and Réunion herons
Rodrigues Owl	*Mascarenotus murivorus*	Mentioned by Leguat and Tafforet
Rodrigues Pigeon	*Columba rodericana*	Bones discovered in 1870s
Rodrigues Starling	*Necropsar rodericanus*	Bones found in 1874 of a 'starling-like' bird
Solitaire	*Pezophaps solitaria*	Accurately described by Leguat and Tafforet

Appendix G
Animals and birds found in Rodrigues today

Animals

Common name	Scientific name	Notes
Mourning Gecko	*Lepidodactylus lugubris*	Probably introduced
Rodrigues Fruit Bat	*Pteropus rodricensis*	Endemic

Birds

Common name	Scientific name	Notes
Bridled Tern	*Sterna anaethetus*	Mainly on Île Cocos
Common Noddy	*Anous stolidus*	Mainly on Île Cocos
Common Tern	*Sterna hirundo*	On mainland coast
Fairy Tern	*Gygis alba*	Mainly on Île Cocos
Lesser Noddy	*Anous tenuirostris*	Mainly on Île Cocos
Little Green Heron	*Butorides striatus*	On shore and estuarine pools
Rodrigues Fody	*Foudia flavicans*	Endemic and threatened
Rodrigues Warbler	*Acrocephalus rodericanus*	Endemic and threatened
Wedge-tailed Shearwater	*Puffinus pacificus*	On mainland coast
White-tailed Tropicbird	*Phaethon lepturus*	Large sea bird, nests in cliffs

Appendix H
Some coastal trees found in Rodrigues

The following are some of the endemic and native trees of Rodrigues, which visitors are now more likely to see being planted in the coastal regions:

Dictyosperma album var. *aureum*
(Palmiste Bon)
This large palm grows to about 8 m in height and gives both plentiful shade and protects the coast. Only two mature individuals exist in Rodrigues. Its heart (where the fronds meet) used to be used in salads, but removing this kills the entire tree. Many have been out-competed by the *Dictyosperma* that was introduced from Mauritius.

Dodonaea viscosa
(Bois de Reinette)
The subspecies native to the Mascarenes is *D. augustifolia*. At one time this plant was considered to be extinct in Rodrigues but a couple of individuals were found on the southern islets. Since then it has been grown successfully from seed. The many roots of this plant help to stabilise the sand and soil.

Gastonia rodriguesiana
(Bois Blanc)
This tree can grow out of limestone rock and is mainly found in the Plaine Corail area. It is currently listed as critically endangered, with less than fifty individuals in the wild. The bark is smooth and the tree can reach up to 12 m in height. It is heterophyllous; the slim leaves of the young plant change to broader green leaves as it matures.

Hyophorbe verschaffeltii
(Palmiste Marron)
This tree is also called the Spindle Palm. It can grow up to 6 m in height and is often cultivated for ornamental purposes: either for gardens or the leaves are used for decoration at festivities. It is considered to be critically endangered with less than sixty individuals in the wild.

Sarcanthemum coronopus
(Oeillet Malabar)
In the 1980s this plant was known only from three islets and one area on the mainland. Île Gombrani is the only place where it occurs naturally today, and there are just thirty individuals growing in the wild. They have spreading roots that are very good at stabilising the sand.

Terminalia bentzoë (subspecies *rodriguensis*)
(Bois Charron)
This endangered tree grows to about 12 m in height. It is heterophyllous; the thin, furry brown leaves change to round, green leaves as the tree matures.

Perhaps the rarest endemic tree now on the island is the Bois Pipe (*Dombeya rodriguesiana*).

Solitaire from an article by R. B. Sharp in *Cassell's Natural History* (1894).

Bibliography

Angier, N. (2002) A new look at the long-lost Dodo and its family tree. *The New York Times*, Thursday 12 March.

Atkinson, G. (1921) A French desert island novel of 1708. *Publications of the Modern Language Association of America*, vol. 36, no. 4, pp. 509–28.

Attenborough, D. (1998) *The Life of Birds*. London: BBC Books.

Balfour, I. B. (1879) Botany of Rodriguez. *Phil. Trans.*, vol. 168, pp. 326–87.

BBC News (2002) *DNA yields dodo family secrets* [online]. Available from: http://news.bbc.co.uk/1/hi/sci/tech/1847431.stm [Accessed December 2006].

Bertuchi, A. J. (1923) *The Island of Rodriguez, a British Colony in the Mascarenhas Group*. London: John Murray.

Bontekoe van Hoorn, W. (1650) *Journael ofte Gedenk waerdige beschrijvinghe van de Oost-Indische Reyse van Willem Ysbrantz Bontekoe van Hoorn*. Amsterdam.

Boudriot, J. (2000) *Cavalier de la Salle. L'Expédition de 1684 – La Belle*. Paris: Éditions ANCRE.

Bour, R. (1984) La répartition géographique des tortues terrestres et d'eau douce aux îles de Maurice et de Rodrigues. *Mauritius Institute Bulletin*, vol. 10, part 1, pp. 75–102.

Buffon, G. L. L. (1770) *Histoire naturelle des oiseaux*. Paris: L'Imprimerie Royale.

Caldwell, W. J. (1875) Notes on the zoology of Rodriguez. *Proceedings of the Zoological Society*, pp. 644–7.

Cantino, A. (1502) *Carta da navigar in le parte de l'Indie*. [Housed at Bibliotheca Estense, Modena].

Carré, A. (1699) *Voyages des Indes Orientales, mêlé de plusieurs histoires curieuses, vol. 2*. Paris: La veuve de Claude Barbin.

Cauche, F. (1651) *Relation du voyage que François Cauche de Rouen a fait à Madagascar, isles adjacentes et coste d'Afrique*. Paris: Augustin Courbé.

Cheke, A. S. (1987) An ecological history of the Mascarene Islands, with particular reference to extinctions and introductions of land vertebrates. In: Diamond, A. W. (ed.) *Studies of Mascarene Island Birds*. Cambridge: Cambridge University Press, pp. 5-89.

Cheke, A. S. (2006) Obituary France Staub (1920-2005). *Ibis*, vol. 148, no. 3, pp. 610.

Clark, G. (1859) A ramble round Mauritius with some excursions in the interior of the island – a familiar description of its fauna and some subjects of its flora by a country school-master. *The Mauritius Register*.

Clusius, C. (1605) *Exoticorum libri decem*. Antwerp.

Corby, T. (1845) [*MS report* dated 30 October. C.O. 167, vol. 262. Public Record Office, London].

Cuvier, G. (1816) *Le règne animal, vol. 3 (Oiseaux)*. Paris: Fortin, Masson et Cie.

Cuvier, G. (1830) *Note sur quelques ossements qui paraissent appartenir au dronte, espèce d'oiseau perdue seulement depuis deux siècles*. Paris.

D'Heguerty, P.-A. (1754) *Memoires de la Société Royale des Sciences et Belle Lettres de Nancy*, vol. 1, p. 29.

Darwin, C. R. (1859) *On the Origin of Species by Means of Natural Selection*. London: John Murray.

Diamond, A. W. *et al.* (1987) *Save the Birds*. Cambridge: Cambridge University Press.

Dubois, le Sieur (1674) *Voyages faits par le Sieur D. B. aux îles Dauphine de Madagascar et Bourbon au Mascarenne les années 1669, 1670, 1671 et 1672*. Paris: Claude Barbin.

Durrell, G. (1977) *Golden Bats and Pink Pigeons*. London: Companion Book Club.

Éditions Le Printemps Ltée (2001) *Tourist Map of Mauritius and Rodrigues*. Mauritius: ELP.

Franklin, B. (1789) [*Codicil from will*, Philadelphia, 23 January].

Fuller, E. (2001) *Extinct Birds*. Cornell, New York: Comstock Publishing Associates.

Fuller, E. (2002) *Dodo – From Extinction to Icon*. London: HarperCollins.

Glendinning, L. (2006) Harriet the tortoise dies aged 175. *Timesonline*, 23 June 2006. Available from: http://www.timesonline.co.uk/tol/news/world/article1083738.ece [Accessed December 2006].

Gomy, Y. (1973) Voyage en île d'amertume. *Info-Nature*, no. 9, pp. 72–99.

Greenway, J. C. (1967) *Extinct and Vanishing Birds of the World*. New York: Dover.

Grihault, A. (2005a) *Dodo – the Bird Behind the Legend*. Cassis, Mauritius: IPC.

Grihault, A. (2005b) *Dodo – the Bird Behind the Legend* [online]. Available from: http://www.dodosite.com [Accessed January 2006].

Grihault, A. (2006) Mare aux Songes. The first ever Dodo bones, a fascinating story. *L'Express*, 12 June.

Grzimek, H. C. B. (1968) *Animal Life Encyclopaedia: Birds*. London: Van Nostrand Reinhold.

Gunther, A. and Newton, E. (1879) The extinct birds of Rodriguez (with the exception of the Solitaire …) *Philosophical Transactions*, vol. 168 (Extra vol.), pp. 423–37.

Hachisuka, M. (1953) *The Dodo and Kindred Birds*. London: H. F. and G. Witherby Ltd.

Halliday, T. (1978) *Vanishing Birds*. New York: Holt, Reinhart and Winston.

Heeringa, K. (1895) De Nederlanders op Mauritius en Madagascar. *De Indische Gids*, 17, pp. 864–92, 1005–36.

Hengst, J. den (2003) *The Dodo - the Bird that Drew the Short Straw*. Marum, Netherlands: Art Revisited.

Herbert, T. (1638) *A Relation of Some Years Travels into Divers Parts of Asia and Africa and the Greater Asia …* London: W. Stansby and J. Bloome.

Hermit (2002) *Mauritius, Hurricanes and the Late 17th Century* [online]. Available from: http://www.churchofvirus.org/archive/0201/0834.html [Accessed September 2006].

Higginson, H. P. (1865) *Reminiscences of life and travel 1859–1872* [personal journal]. New Zealand.

Hume, J. P. (2003) *The Penguin of Mauritius – the Real Facts about the Dodo*. [Unpublished thesis].

Hume, J. P. and Cheke, A. S. (2004) The White Dodo of Réunion Island: unravelling a scientific and historical myth. *Archives of Natural History*, vol. 31, no. 1, pp. 57–79.

Jenner, G. (1871) *Report on the Remains of the Solitaire in the Caverns of Rodrigues in the Months of January and February, 1871*. [MS copy in the possession of the Balfour Library, University of Cambridge].

Keating, H. (1809) [*Letter* dated 20.8.09, MS, fol. 9464, Bombay Castle Records, India Office, London].

Kinn, F. (1904) *Wild Animals of Yesterday and Today*. London: S. W. Partridge.

Knip, P. (1811) *Les pigeons*. Paris: Chez Mme Knip.

L'Estrange, H. (1638) Manuscript – Sloane MSS. In: Wilkin, S (ed.) (1839) *Sir Thomas Brown's Works*, 4 vols. London.

Labistour (Capitaine) (1786) [Entry for Ship *Chaloupe du Brab* into Mauritius from Rodrigues].

Lagesse, J. (2004) *The oldest Dodo in the World*. Mauritius: Pierre Lagesse.

Le Monnier, P.-C. (1776) Constellation du Solitaire. *Mémoires de l'Académie Royale des Sciences*.

Leguat, F. (1708) *Voyage et avantures de François Leguat et de ses compagnons en deux isles désertes des Indes Orientales*. London: David Mortier.

Livezy, B. C. (1993) An ecomorphological review of the Dodo (*R. cucullatus*) and the Solitaire (*Pezophaps solitaria*), flightless Columbiformes of the Mascarene Islands. *Journal of Zoology*, 230, pp. 247–92.

Lucas, F. A. (1893) The weapons and wings of birds. *Report of the U.S. National Museum for 1893*, pp. 653–63.

Marragon, P. (1795) *Memoire sur l'Isle Rodrigue*. [Mauritius Archives, TB 5/2. 8 July, p. 17].

Millard, C. (2004) *Rodrigues, the First Settlement – 1691*. Port Louis, Mauritius: Christian Heritage Ministries.

Milne-Edwards, A. (1866-73) *Recherches sur la faune ornithologique tiente des Îles Mascareignes et de Madagascar*. Paris.

Milne-Edwards, A. (1896) Sur les resemblances qui existent entre la faune des Îles Mascareignes et celle de certaines îles de l'Ocean Pacifique Australe. *Annales des Sciences Naturelles*, 8e série, vol. 2.

Moreau, C. (1999) *Le Solitaire de L'Île Rodrigues*. Mauritius: Collection Le Solitaire.

Mourer-Chauviré, C., Bour, R. and Ribes, S. (1995) Position systémique du Solitaire de la Réunion: nouvelle interprétation basée sur les restes fossiles et les récits des anciens voyageurs. *Comptes Rendus de l'Académie des Sciences (Paris)* série II a, 320, pp. 1125–31.

Mundy, P. (1628) *Peter Mundy's Journal*. [Reprinted by the Hakluyt Society, 1914].

Newton, A. (1867) Discovery of *Didus solitarius. Transactions of the Royal Society of Arts and Sciences, Mauritius*. [Revised and reprinted in *Mauritius Almanac* (1869), pp. 41–6].

Newton, A. and Gadow, H. (1896) *A Dictionary of Birds*. London: A. & C. Black.

Newton, A. and Newton, E. (1868) On the osteology of the Solitaire or Didine Bird of the island of Rodriguez (Gmel). *Proceedings of the Royal Society,* 103, pp. 428–33.

Newton, E. (1865) Notes on a visit to the island of Rodriguez. *Ibis,* 2nd series, vol. 1, pp. 146-54.

Newton, E. (1875) Additional evidence as to the original fauna of Rodriguez. *Proceedings of the Zoological Society of London,* pp. 39–43.

North-Coombes, A. (1971) *The Island of Rodrigues*. Mauritius: [the author].

North-Coombes, A. (1979) *The Vindication of François Leguat*. Mauritius: [the author].

Oliver, S. P. (ed.) (1891) *The Voyage of François Leguat of Bresse to Rodriguez, Mauritius, Java, and the Cape of Good Hope (Vols. 1 & 2)*. London: The Hakluyt Society.

Oustalet, E. (1896) Notice sur la faune ornithologique ancienne et moderne des îles Mascareignes et en particulier de l'île Maurice d'aprés des documents inédits. *Ann. Sci. Nat. Zool. Et Paleo,* vol. 3, no. 2.

Owen, R. (1866) *Memoir on the Dodo*. London: Taylor and Francis.

Owen, R. (1878) On the Solitaire. *Annals and Magazine of Natural History*, vol. I, no. 1, pp. 87–98.

Patnaik, N. (1993) *The Garden of Life*. London: HarperCollins.

Pingré, A. G. (1761) *Relation de mon voyage de Paris à l'Isle Rodrigue,* MS 1803 and *A narrative of the island with observations thereon,* MS 1804 (Library Sainte Geneviève); *Journal de Voyage Pingré,* MS 2537 (Bibliothèque du Service Hydrographique de la Marine).

Pinto-Correia, C. (2003) *Return of the Crazy Bird. The Sad, Strange Tale of the Dodo*. New York: Copernicus Books.

Pitot, A. (1914) Extinct birds of the Mascarene Islands. In: Macmillan, A. (ed.) *Mauritius Illustrated*. London: W. H. & L. Collingridge.

Pritchard, P. C. H. (2005) *Tortoise Life* [online]. Available from: http://www.slowcoach.org.uk/articles/artic04/artcl_04.html [Accessed April 2006].

Reinhardt, J. (1842) Nojere Oplysning om det I Kjoben-havn funde Drontehoved. *Naturh Tidkr,* vol 4.

Rijsdijk, K. F. *et al.* (2005) *Hundreds of Rare Dodo Bones Discovered*. [Press release].

Rijsdijk, K. F. *et al.* (2006) Findings from the Dodo Research Programme digs at Mare aux Songes in 2005 and 2006. [Paper presented to the *Mauritian-European Dodo Research Programme* on 29 November 2006].

Rothschild, W. (1907) *Extinct Birds*. London: Hutchinson.

Rowley, G. D. (1877) On the extinct birds of the Mascarene Islands. In: Rowley, G. D. (ed.) *Ornithological Miscellany*. London: Trübnor and Co., pp. 123–33.

Schlegel, H. (1866) On some extinct gigantic birds of the Mascarene Islands. *Ibis,* vol. 2.

Shapiro, B. *et al.* (2002) Flight of the Dodo. *Science,* vol. 295, no. 5560, pp. 1683.

Sharp. R. B. (1894) *Aves*. In: Duncan, P. M. (ed.) *Cassell's Natural History*. London: Cassell and Company.

Shoals Rodrigues (2004) *The History of Rodrigues* [online]. Available from:
http://www.shoals-rodrigues.org/island7a.html [Accessed January 2006].

Silverberg, R. (1969) *The Auk, the Dodo and the Oryx; Vanished and Vanishing Creatures.*
London: World's Work.

Sinclair, J. M. (1998) *Collins English Dictionary*. Glasgow: HarperCollins.

Slater, H. H. (1879) Observations on the bone caves of Rodriguez. *Philosophical Transactions of the Royal
Society of London*, series B, 168, pp. 420–2.

Spooner, P. (1992) Museum of the mind. In: Nicholson, G. (2004) *The Hollywood Dodo*. London:
Serpent's Tail.

Staub, F. (1996) Dodos and Solitaires, myths and reality. *Proceedings of the Royal Society of Arts and
Sciences, Mauritius*, vol. 6, pp. 89–122.

Staub, F. (2000) New hypothesis on the Dodo's true morphology from an ecological consideration of
its available diet. Paper given at a joint seminar on *Globalisation and the South-West Indian
Ocean*, 21–23 September 1998, Mauritius.

Storer, R. W. (2005) *Independent Evolution of the Dodo and the Solitaire* [online].
http://elibrary.unm.edu/sora/Auk/v087n02/p0369-p0370.pdf [Accessed November 2006].

Strickland, H. E. and Melville, A. G. (1848) *The Dodo and its Kindred*. London: Reeve,
Benham and Reeve.

Tafforet, J. (1726) Relation de l'isle Rodrigue. In: Milne-Edwards, A. (1875). Nouveaux documents sur
l'époque de la disparition de la faune ancienne de l'île Rodrigue. *Ann. Sd. Nat. Zoologie*,
6th series, vol. 2.

Tatton, J. (1625) A journal of a voyage made by the Pearl to the East Indies wherein went as Captain
Master Samuel Castleton of London, and Captain George Bathurst as Lieutenant.
In: Purchas, S. (1905) *Purchas His Pilgrimes*. London.

Thompson, J. V. (1829) Contributions towards the natural history of the Dodo (*Didus ineptus*, Lin).
Magazine of Natural History, 2, pp. 442–8.

Valleau, A. (1692) Interrogatoire d'Antoine Valleau. *Archives Nationale*. sect. Colonies, C3,
pièce 4. Paris.

Valledor de Lozoya, A. (2003) An unnoticed painting of a white dodo. *Journal of the History of
Collections*, vol. 15, no. 2, pp. 201–10.

van Heygen, E. (2004a) *The Dodo's Stepping Stones the Same as Phelsuma?* [online]. Available from:
http://www.phelsumania.com/public/articles/biogeography_dodo.html [Accessed
September 2005].

van Heygen, E. (2004b) *Rodrigues* [online]. Available from:
http://www.phelsumania.com/public/biogeography/mascarenes/rodrigues.html
[Accessed September 2005].

Vince, G. (2006) Bird extinction rates far worse than realised. *New Scientist.com news service* [online].
Available from: http://www.newscientist.com/article/dn9472-bird-extinction-rates-far-
worse-than realised.html [Accessed December 2006].

Vinson, J. and Vinson, J. M. (1969) The Saurian fauna of the Mascarene Islands. *Mauritius Institute
Bulletin*, 6, pp. 203–320.

Wood, J. G. (1897) *Illustrated Natural History*. Philadelphia: Henry Altemus.

Index